A Guide To
Learning
Statistics

Robert F. Herrmann

A Guide To
Learning
Statistics

••••••••••••••••••••••••••••••

Robert F. Herrmann

Elmhurst College, Illinois

The McGraw-Hill Companies, Inc.

New York   St. Louis   San Francisco   Auckland
Bogotá   Caracas   Lisbon   London   Madrid   Mexico City
Milan   Montreal   New Delhi   San Juan   Singapore
Sydney   Tokyo   Toronto

McGraw-Hill

A Division of The McGraw·Hill Companies

A Guide To Learning Statistics

1 2 3 4 5 6 7 8 9 0   DOC/DOC   9 9 8 7 6

ISBN 0-07-028466-0

The editor was Maggie Lanzilla
the production supervisor was Kathryn Porzio.
R. R. Donnelley & Sons Company was printer and binder.

Library of Congress Cataloging-in-Publication Data is available.
LC Card#96-75378

1. Introduction

How to Use This Guide

Audience

Use this guide if you are taking an elementary course in statistics, whether the course is from mathematics, business, or any other department. The elementary concepts presented in this guide are universally applicable.

Organization

This guide covers topics most often covered in a one-semester elementary statistics course. It contains explanations, procedures, examples, and exercises in a conversational style that complements any textbook. Explanations are brief and succinct. Procedures are written in a "step-action" fashion for ease of use and reference. Examples are simple yet contain all the points of the explanations. Exercises are meant to bridge this guide to the student's textbook.

Special Symbols

The following visual cues are found throughout this guide. They highlight items of which you should be aware.

✔ represents a "sanity check" that you can use to make sure you are solving the problem correctly.

? represents a simple question that you can answer as you read the study guide to assure that you understand the basic concepts. Brief answers to these questions can be found in the back of this study guide.

! represents an important idea or a helpful hint.

◆ represents a formula or rule to be used in solving problems.

Mathematical Conventions

- Statistics texts differ in the usage of variables to represent various concepts. For example, one text might use a lowercase letter to represent a random variable. Another text might use the same letter in uppercase for this idea. Still another text might use a different letter altogether to represent a random variable. Though your text may vary from this guide in these matters, the concepts the variables represent remain the same.

- Generally, this study guide uses summations without indexes. For example, $\sum_{i=1}^{n} x_i$ is

  written $\sum x$.

Basic Terms

Certain terms form the basis of the vocabulary of statistics. Some of these terms appear below. They and their meanings occur throughout this guide.

Data

Data are measurements of a situation under consideration. Synonyms for data are scores, measurements, and observations.

Example 1

The heights of all adults in the world might be recorded. These recorded heights make up our data.

Areas of Statistics

There are two area of **statistics** as part of mathematics: 1) descriptive, and 2) inferential.

Descriptive Statistics

Descriptive statistics involve the collection, description, and summarization of data. The first part of this guide deals with descriptive statistics.

Statistics	
Descriptive	Inferential

Inferential Statistics

Inferential statistics involve drawing a conclusion based on the data. The last part of this guide deals with descriptive statistics.

Population

A **population** is the collection of *all* possible data for the situation under consideration. In terms of algebra, a population is the universe set of a **Venn diagram** as shown here.

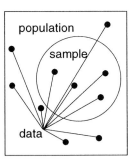

Example 2

The population from example 1 is made up of heights of all adults in the world.

Sample

A **random sample** is a collection of *some* data taken from a given population. The data is chosen at random to get a fair selection from the population. In terms of algebra, a sample is a subset of the population.

Example 3

A sample from the population in example 2 might include heights of 10,000 adults randomly chosen from around the world.

Parameter

A **parameter** is a number describing a characteristic of a *population*.

Example 4

The average height of all adults in the world might be 5 feet 8 inches. This average is a parameter of the population.

! A parameter does not necessarily appear as a measurement in the population. In example 4, there might not be any adults in the world measuring exactly 5 feet 8 inches.

Statistic

A **statistic** is a number describing a characteristic of a *sample*.

Example 5

Variability measures how "spread out" data are. Suppose the variability of the height measurements of the adults in example 3 is 1.2 feet. Variability is a statistic of that sample.

Deduction

Deductive reasoning draws with certainty a specific conclusion from general assumptions.

Example 6

General assumptions: All birds have feathers. A robin is a bird.

Specific conclusion: We are certain that a robin has feathers.

Induction

Inductive reasoning draws, correctly or incorrectly, a general conclusion from specific observations. The certainty of this conclusion is stated with the help of **probability**. The probability of the conclusion is determined through methods that we study later in this guide.

$$\text{General} \xrightarrow{\quad\text{Deduction}\quad} \text{Specific}$$
$$\text{General} \xleftarrow{\quad\text{Induction}\quad} \text{Specific}$$

Example 7

Specific observation: It has rained on the past seven Mondays.

General conclusion: It will rain this Monday as well.
Note that we can include a probability with the statement of this
conclusion. For instance, after studying the data, we might say that
there is a 75% chance that it will rain this Monday as well.

2. Describing Data

Tabulating Data

Grouped and Ungrouped Data

Summarize your data either as ungrouped or grouped. **Ungrouped** data are the actual raw values. For example, 2, 2, 5, 7, 8, 9 are ungrouped data. A benefit to ungrouped data is that the actual values are available for further study. A drawback is that too many ungrouped data are hard to analyze except with a computer.

Data becomes **grouped** when you bunch them into informative intervals. For example, the data 2, 2, 5, 7, 8, 9 can be grouped as 0-2, 3-5, 6-8, 9-11. A benefit to grouped data is that a large number of data can be summarized into a smaller, more workable number of groups. A drawback to grouped data is that, unless you save the actual values after grouping, they are not available for further analysis.

Distributions

Distributions are tables used to summarize data. Present your data either as ungrouped data or as grouped data broken into **class intervals**. Each class interval is non-overlapping and contiguous.

! Determining the number of class intervals for your distribution is a subjective process. Choose a number that best displays information about your data, that shows the general trends of your data, and that does not mask concentrations or "gaps" in your data. The best way to learn how to choose this "magic" number is through practice.

! For a rule of thumb, use from 5 to 20 class intervals. As you can imagine, varying the number of class intervals will likely vary the way the data are represented in the table.

Example

The remainder of this section of the study guide uses the following example: Suppose IQ scores for 150 third-grade students run from 86 to 132. You want to summarize the data using five class intervals.

Constructing Class Intervals

To construct the class intervals for this example, first determine the **range** of the data:

                 range = highest value – lowest value

For this example,

range = 132 – 86 = 46

To determine the **class width**, use:

✦
$$\text{class width} = \frac{\text{range}}{\text{number of classes}}, \text{ rounded up}$$

In the table below, class width $= \dfrac{132-86}{5} = \dfrac{46}{5} = 9.2$, rounded up to 10.

! Generally, in any distribution, the width of all class intervals is *constant.*

With the range and class width determined, construct the class intervals. In this example, the minimum IQ is 86. Starting the first class interval at 85 with a class width of 10 provides "comfortable" numbers with which to work. Remember, this is a subjective process.

Class Interval
85-94
95-104
105-114
115-124
125-134

✔ Class width should be
    constant.

! Start the first class interval at a number less than or equal to the minimum value.

Parts of Class Intervals

Look at the class intervals in the table. The **upper class limits** are 94, 104, 114, …. The **lower class limits** are 85, 95, 105, …. **Class boundaries** are the averages of an upper class limit and the succeeding lower class limit. In this example, the class boundaries are 84.5, 94.5, 104.5, …. **Class marks (class midpoints)** are 89.5, 99.5, 109.5, ….

? What is the class boundary between an upper class limit of 0.05 and a succeeding lower class limit of 0.06?

Types of Distributions

! In this study guide, we have these definitions:

$n =$ the number of data in each class interval (frequency)

$N =$ the total number of data (total frequency)

Note that uppercase and lowercase letters have different meanings. You will see this throughout your study of statistics.

There are four basic distributions: frequency, cumulative frequency, relative frequency, and cumulative relative frequency. We study each next.

A **frequency distribution** is made up of the class intervals and a second column containing tallies of data in each interval. These tallies are called the **frequencies**. From the frequency distribution for this example, we see that there are 34 values in the 95-104 class (i.e., a frequency of 34).

Class Interval	Frequency (n)
85-94	20
95-104	34
105-114	49
115-124	35
125-134	12
	$N = 150$

✔ Total of this column always equals N.

?　　　　What is the frequency of values in the 115-124 class?

A **cumulative frequency distribution** builds on the frequency distribution. A third column accumulates the frequencies from the second column.

Class Interval	Frequency (n)	Cumulative Frequency
85-94	20	20
95-104	34	20 + 34 = 54
105-114	49	103
115-124	35	138
125-134	12	150
	$N = 150$	

✔ Last entry in this column always equals N.

If, to a frequency distribution, we add a column that contains a probability (percentage) that a value occurs in each class, we create a **relative frequency distribution**. In this example, there is a 32.7% chance that a value falls into the 105-114 class.

Class Interval	Frequency (n)	Relative Frequency (n/N)
85-94	20	20/150 = 0.133
95-104	34	34/150 = 0.227
105-114	49	0.327
115-124	35	0.233
125-134	12	0.080
	$N = 150$	1.000

✔ Total of this column always equals 1.

If we accumulate the relative frequencies from a relative frequency distribution, we have a **cumulative relative frequency distribution**. For example, in the table, 68.7% of the values are found between 85 and 114.

Class Interval	Frequency (n)	Relative Frequency (n/N)	Cumulative Relative Frequency
85-94	20	20/150 = 0.133	0.133
95-104	34	34/150 = 0.227	0.133 + 0.227 = 0.360
105-114	49	0.327	0.687
115-124	35	0.233	0.920
125-134	12	0.080	1.000
	$N = 150$	1.000	

✔ Last entry in this column always equals 1.

? In the table, what percentage of values lie between 85 and 124?

Exercises

1. Suppose you are studying data in which the lowest value is 31 and the highest value is 97. You decide you want 8 class intervals.
    a) What should your class width be?
    b) Construct the class intervals.
    c) What might the lower class limits be? What might the upper class limits be? (Remember that this is a subjective process and that there are many answers to the last two questions.)

2.* The following represent the number of customers ordering "secret sauce" on their hamburgers at lunch during each of 30 consecutive days in a survey by the owner of a local restaurant chain:

    7, 4, 8, 5, 2, 7, 7, 4, 6, 9, 5, 3, 4, 6, 7, 9, 5, 6, 8, 7, 5, 8, 6, 6, 2, 3, 1, 1, 6, 9

    Using 5 class intervals of width 2:
    a) Construct a frequency distribution.
    b) Construct a cumulative frequency distribution.
    c) Construct a relative frequency distribution.
    d) Construct a cumulative relative frequency distribution.

3.* GPAs for 50 statistics students are:

    2.6, 3.0, 2.4, 2.8, 3.1, 3.4, 2.7, 2.9, 2.5, 2.2, 3.7, 3.3, 2.7, 2.2, 2.1, 2.7, 2.8, 2.4, 3.1, 3.7, 2.0, 2.5, 2.7, 2.8, 3.8, 3.1, 2.6, 2.4, 2.5, 2.2, 3.5, 3.0, 2.3, 2.9, 2.9, 2.6, 2.1, 3.7, 2.0, 2.6, 2.4, 2.3, 3.5, 2.7, 2.8, 3.0, 2.9, 2.6, 2.4, 3.0

    Using 6 class intervals:
    a) Construct a frequency distribution.
    b) Construct a relative frequency distribution.

    * Save your answers for exercises in the following section.

Graphing Data

Pie Charts

The entire pie (circle) represents all, or 100%, of the data. Each slice of the pie (sector) represents a class or subdivision of the data. Use a relative frequency distribution to create these sectors.

✦   Number of degrees in the sector = relative frequency times 360

For example, the green sector contains $0.23 \cdot 360 = 82.8°$.

Color	Frequency	Rel. Freq.
Blue	88	0.44
Green	46	0.23
Orange	24	0.12
Red	6	0.03
Yellow	36	0.18

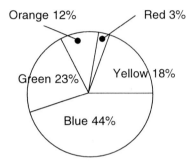

?   How are relative frequencies calculated?

Bar Charts

This type of graph shows frequencies for *ungrouped* data. Use a frequency distribution to create this graph.

Ounces	Frequency
1	29
2	51
3	59
4	45
5	16

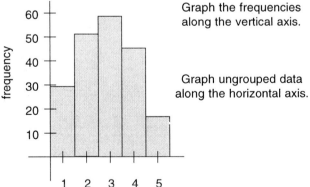

Graph the frequencies along the vertical axis.

Graph ungrouped data along the horizontal axis.

Histograms and Relative Histograms

A histogram shows frequencies for *grouped* data. A relative histogram shows *relative frequencies* for *grouped* data. Use a frequency distribution or a relative frequency distribution to create this graph.

? What is the difference between grouped and ungrouped data?

Ounces	Frequency
1-2	29
3-4	51
5-6	59
7-8	45
9-10	16

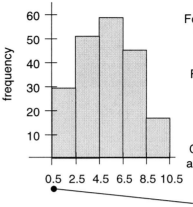

For histograms, graph the frequencies along the vertical axis.

For *relative* histograms, graph the *relative* frequencies along the vertical axis.

Graph class boundaries along the horizontal axis. Start with the leftmost class boundary at the origin.

Frequency Polygons

Use the frequency polygon to graph a frequency distribution.

Ounces	Frequency
1-2	29
3-4	51
5-6	59
7-8	45
9-10	16

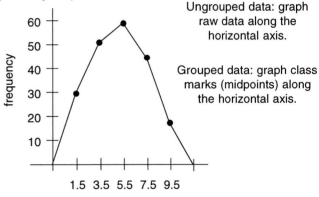

Ungrouped data: graph raw data along the horizontal axis.

Grouped data: graph class marks (midpoints) along the horizontal axis.

Relative Frequency Polygons

Use the relative frequency polygon to graph a relative frequency distribution.

Ounces	Rel. Freq.
1-2	0.145
3-4	0.255
5-6	0.295
7-8	0.225
9-10	0.080

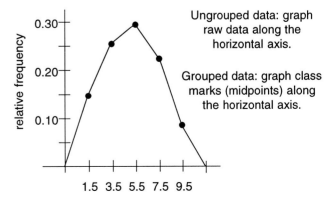

Ungrouped data: graph raw data along the horizontal axis.

Grouped data: graph class marks (midpoints) along the horizontal axis.

Ogives (Cumulative Frequency Polygons)

To picture the data from a cumulative frequency distribution, use an ogive (pronounced oh´-jive). An ogive is also called a cumulative frequency polygon.

✔ The graph rises from left to right as you would expect from accumulating frequencies.

Ounces	Cum. Freq.
1-2	29
3-4	80
5-6	139
7-8	184
9-10	200

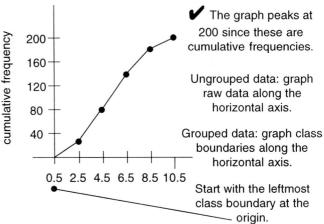

✔ The graph peaks at 200 since these are cumulative frequencies.

Ungrouped data: graph raw data along the horizontal axis.

Grouped data: graph class boundaries along the horizontal axis.

Start with the leftmost class boundary at the origin.

Exercises

1. The marital status of 25 seminar participants was recorded.
    a) Construct a pie chart.
    b) Construct a bar chart.

Marital Status	Frequency
Single	10
Married	8
Divorced	5
Widowed	2

2. The following represent the number of customers ordering "secret sauce" on their hamburgers at lunch during each of 30 consecutive days in a survey by the owner of a local restaurant chain:
    7, 4, 8, 5, 2, 7, 7, 4, 6, 9, 5, 3, 4, 6, 7, 9, 5, 6, 8, 7, 5, 8, 6, 6, 2, 3, 1, 1, 6, 9
    Use 5 class intervals to:
    a) Construct a relative histogram.
    b) Construct a frequency polygon.
    c) Construct an ogive.

3. GPAs for 50 statistics students are:
    2.6, 3.0, 2.4, 2.8, 3.1, 3.4, 2.7, 2.9, 2.5, 2.2, 3.7, 3.3, 2.7, 2.2, 2.1, 2.7, 2.8, 2.4, 3.1, 3.7,
    2.0, 2.5, 2.7, 2.8, 3.8, 3.1, 2.6, 2.4, 2.5, 2.2, 3.5, 3.0, 2.3, 2.9, 2.9, 2.6, 2.1, 3.7, 2.0, 2.6,
    2.4, 2.3, 3.5, 2.7, 2.8, 3.0, 2.9, 2.6, 2.4, 3.0
    Use 6 class intervals to:
    a) Construct a relative frequency polygon.
    b) Construct a relative histogram.

Stem-and-Leaf Plots and Boxplots

Stem-and-Leaf Plots

A **stem-and-leaf plot** is a blend of a table and a graph. It is a table because the raw data is grouped into classes. It is a graph because it mimics a histogram.

? Describe a histogram.

Example 1

The stem-and-leaf plot for temperature readings in a laboratory refrigeration unit are shown below. After arranging the data in ascending order, each reading is separated into two parts: a **stem** and a **leaf**.

Data	Stem	Leaves
37.6	37.	6
38.0, 38.0, 38.9	38.	009
39.0, 39.0, 39.2, 39.4, 39.8	39.	00248
40.0, 40.6, 40.7, 40.9	40.	0679
41.1, 41.3	41.	13
42.7	42.	7

Stems are the leftmost digits of the data. They represent an informative grouping of the data.

Leaves are the rightmost digits of the data.

! Choose the stem in the same way you choose class intervals for a frequency distribution, so that information about your data is shown.

✔ Turned sideways, the "leaves" column has the same profile as a histogram of the data.

Example 2

Construct the stem-and-leaf plot for these broad-jump readings (in feet):

12.85, 12.88, 13.01, 13.03, 13.03, 13.03, 13.06, 13.09, 13.10, 13.12, 13.13, 13.20

Notice that the data is arranged from least to greatest.

Stem	Leaves
12.8	58
12.9	–
13.0	133369
13.1	023
13.2	0

Nothing entered indicates no data for this stem.

Boxplot

A **boxplot** shows how the values are dispersed. Use the boxplot to uncover data that might have been collected incorrectly.

5-Number Summary

A **5-number summary** is the basis of the boxplot. It consists of the:

*   Minimum value in the data.

*   Maximum value in the data.

*   Median – If there are an odd number of values, the median is the value that divides the data into two equal parts. If there are an even number of values, the median is the average of the two middle values. For example, the median for 2, 2, 3, 4, 5, 7, 8 is 4 since there are an odd number of values, and 4 divides the data into two equal parts of 3 values each. Suppose the data is 2, 2, 3, 4, 5, 7. The median is 3.5, since there are an even number of values, and 3.5 is the average of 3 and 4.

*   Lower hinge – the median of values from the minimum value to the overall median.

*   Upper hinge – the median of values from the overall median to the maximum value.

Constructing a Boxplot

After arranging the data from least to greatest, identify the 5-number summary. Plot these five numbers on a number line. Draw a box from the lower hinge to the upper hinge including the median.

Example 1

Suppose after arranging the data from least to greatest, you identify the 5-number summary for 6.1, 8.2, 11.8, 16.4, 20.0, 23.6, 25.9 as:

minimum value = 6.1   maximum value = 25.9   median = 16.4

The lower hinge is the median of the bottom four data. The upper hinge is the median of the top four data. Since there is an even number of data in each of these groups, we have:

$$\text{lower hinge} = \frac{8.2+11.8}{2} = 10.0 \qquad \text{upper hinge} = \frac{20.0+23.6}{2} = 21.8$$

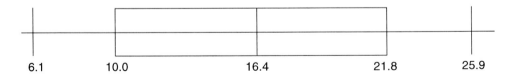

6.1   10.0   16.4   21.8   25.9

The boxplot divides the data into four sections. These four sections appear fairly equal in size. Even dispersion like this indicates reasonably collected data.

Example 2

If the four sections do not appear equal, some raw data might have been collected incorrectly. Suppose a boxplot appears as below. An errant value, or values, might exist between the upper hinge and the maximum value.

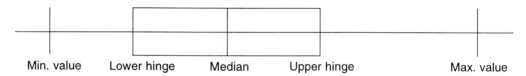

| Min. value | Lower hinge | Median | Upper hinge | Max. value |

! In the case of an uneven boxplot, review how the data in the disproportionate area was collected and recorded.

Exercises

1. Construct a stem-and-leaf plot for the number of municipal library patrons recorded on the hour during a recent weekend:
    136, 123, 96, 110, 93, 85, 105, 121, 117, 85, 120, 91, 114, 130

2. Reconstruct the original data from the following stem-and-leaf plot.

Stem	Leaves
4.2	34568
4.3	4457
4.4	269
4.5	–
4.6	0

3. Find the median of the data in exercise 1 above.

4. Suppose the second 85 of the data in exercise 1 was omitted. Find the median of the resulting data:
    136, 123, 96, 110, 93, 85, 105, 121, 117, 120, 91, 114, 130

5. List the 5-number summary for the data in exercise 1 above. Construct a boxplot for the data.

Describing Paired Data

Paired Data: An Example

Often, a study is concerned with more than one data measurement from a single subject. Suppose the height and weight of students in a statistics class are as follows.

Student	Height (in.)	Weight (lbs.)
1	62	105
2	63	110
3	66	124
4	66	138
5	67	145
6	67	180
7	68	157
8	68	175
9	70	210
10	71	177

The height and weight of each student are called **paired data** since both measurements come from the same subject.

Scatter Diagram: An Example

To create a picture of the study discussed above, represent the paired data as ordered pairs, (x,y). In each ordered pair, x is the student's height and y is the student's weight. The collection of all such ordered pairs from the students in the study can be graphed in a **scatter diagram**.

A scatter diagram is a graph of the paired data (x,y) with x on the horizontal axis and y on the vertical axis.

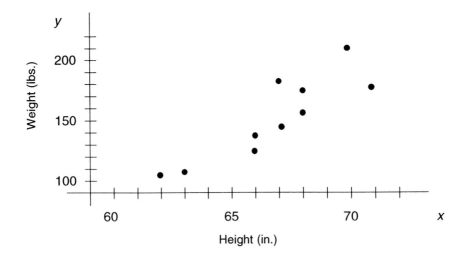

Scatter Diagram: Analysis

A scatter diagram can indicate a relationship between paired data. Plotted points rising as you move from the left to the right across the graph indicate a **positive relationship** between the data. That is, as one measurement increases, so does the other. Consider the previous example. Generally, as height increases, so does weight. Thus, there is a positive relationship between height and weight.

Points dropping as you move left to right indicate a **negative relationship**. For example, as altitude increases, temperature decreases.

Points with no apparent pattern on the scatter diagram show **no relationship** between the paired data. For example, suppose we compare the annual stock price of a candy company during the last decade and the number of annual worldwide volcano eruptions during the same period. We expect no apparent pattern in the scatter diagram and no relationship between these two quantities.

Measuring the strength of the data's relationship is called **correlation**, and is discussed later in this study guide.

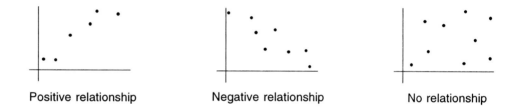

Positive relationship   Negative relationship   No relationship

Exercises

Construct a scatter diagram for each of the following sets of paired data. What relationship, if any, does the scatter diagram reveal?

1.

# of Cars	Mortality Rate per 1000
50,000	4.7
100,000	9.3
150,000	15.6
200,000	25.0

2.

# of Cars	Avg. # of Passengers
50,000	2.9
100,000	3.2
150,000	2.3
200,000	3.0

3.

% of Passengers Using Seatbelts	Mortality Rate per 1000
20	23.6
40	17.2
60	12.8
80	5.2

Solutions to Exercises

Tabulating Data

1. a)   Class width $= \dfrac{97-31}{8} = 8.25$, rounded up to 9

    b)   A possible answer is:
         30-38
         39-47
         48-56
         57-65
         66-74
         75-83
         84-92
         93-101

    c)   Lower class limits appear to the left of the hyphen in b). Upper class limits appear to the right of the hyphen.

2. a) - d) Following is a possible answer, depending on your choice of class intervals.

Orders	Frequency (n)	Cumulative Frequency	Relative Frequency (n/N)	Cumulative Relative Frequency
0-1	2	2	0.067	0.067
2-3	4	6	0.133	0.200
4-5	7	13	0.233	0.433
6-7	11	24	0.367	0.800
8-9	6	30	0.200	1.000
	$N = 30$		1.000	

3. a) - b) Following is a possible answer, depending on your choice of class intervals.

GPA	Frequency (n)	Relative Frequency (n/N)
1.7-2.0	2	0.04
2.1-2.4	12	0.24
2.5-2.8	17	0.34
2.9-3.2	11	0.22
3.3-3.6	4	0.08
3.7-4.0	4	0.08
	$N = 50$	1.000

Graphing Data

1. a)

Status	Frequency	Rel. Freq.	Rel. Freq x 360 Degrees
Single	10	0.40	144.0
Married	8	0.32	115.2
Divorced	5	0.20	72.0
Widowed	2	0.08	28.8
	25	1.00	360.0

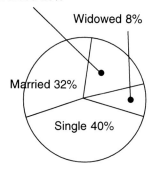

b)

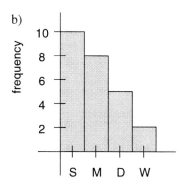

2. Your answers might differ slightly from these, depending on your choice of class limits.

a)

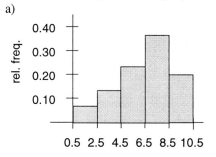

b)

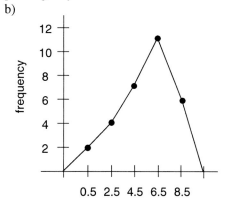

c)

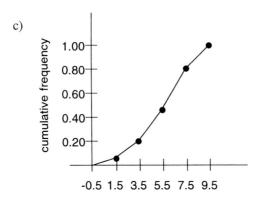

3. Your answers might differ slightly from these, depending on your choice of class limits.

a)

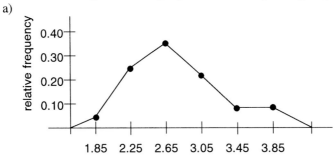

b)

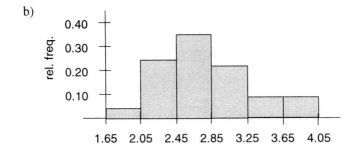

Stem-and-Leaf Plots and Boxplots

1.

Stem	Leaves
8	55
9	136
10	5
11	047
12	013
13	06

2. 4.23, 4.24, 4.25, 4.26, 4.28, 4.34, 4.34, 4.35, 4.37, 4.42, 4.46, 4.49, 4.60

3. Since there are an even number of scores, the median is the average of the 7th and 8th scores, i.e., 110 and 114. So, the median is 112.

4. Omitting the second score of 85 results in an odd number of scores. So, the median is the seventh score, 114 (six values above and below).

5. Minimum score = 85, lower hinge = 93 (3 values above and below)

$$\text{median} = \frac{110+114}{2} = 112$$

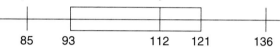

85   93   112   121   136

upper hinge = 121 (3 values above and below), maximum score = 136

Describing Paired Data

1. Data has a positive relationship.

2. Data has no relationship.

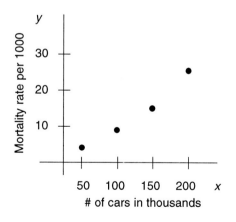

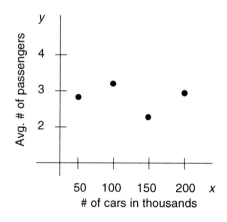

3. Data has a negative relationship.

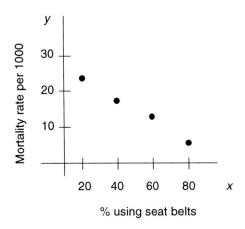

% using seat belts

Answers to ?

Page	Answer
2.2	0.055
2.3	35
2.4	92%
2.6	class frequency divided by total frequency, that is, n/N
2.7	see Grouped and Ungrouped Data, page 2.1
2.9	see Histograms and Relative Histograms, page 2.6

3. Summarizing Data

Central Tendency

Overview

A measure of central tendency (average) gives us a value that tends to be in the center of the rest of the data. We will study *five* ways of measuring central tendency.

? What is the difference between grouped and ungrouped data?

Mean for Ungrouped Data

The (arithmetic) **mean for *ungrouped* data** is:

$$\bar{x} = \frac{x_1 + x_2 + \ldots + x_n}{n} = \frac{\Sigma x}{n}$$

Sum of sample data

Number of sample data

Note: $\bar{x}$ is the sample mean.
μ is the population mean.

? What is the difference between a parameter and a statistic?

Example 1

Find the mean of these data: $x_1 = 7$, $x_2 = 4$, $x_3 = 2$, $x_4 = 3$.

There are four data so $n = 4$ and $\bar{x} = \dfrac{x_1 + x_2 + x_3 + x_4}{4} = \dfrac{7+4+2+3}{4} = \dfrac{16}{4} = 4$

Mean for Grouped Data

The **mean for *grouped* data** is:

$$\bar{x} = \frac{\Sigma fx}{\Sigma f} = \frac{\Sigma fx}{n}$$

where x = class mark, f = class frequency, and n = number of data.

Example 2

Find the mean of these grouped data.

x = Height in centimeters	150-159	160-169	170-179	180-189	190-199
f = # of scores with height x	1	2	4	3	2

$$\bar{x} = \frac{\Sigma fx}{\Sigma f} = \frac{1(154.5) + 2(164.5) + 4(174.5) + 3(184.5) + 2(194.5)}{1+2+4+3+2} = 177$$

Median

The **median** lies in the middle of the sample (population) when data are arranged in order of magnitude. With an *odd number of data*, the median is the middle value. With an *even number of data*, the median is the mean of the two middle values.

Example 3

Arrange the odd number of data in order of magnitude:

7, 2, 5, 3, 4 becomes 7, 5, **4**, 3, 2. So, the median is 4.

Example 4

Arrange the even number of data in order of magnitude:

7, 2, 3, 4 becomes 7, **4**, **3**, 2. So, the median is (4 + 3)/2 = 3.5.

Mode

The **mode** is the one score appearing most frequently. There is *no mode* if each value occurs only once. There can be *more than one mode*, since different values that occur with the same and greatest frequency are all considered modes.

Example 5

7, 7, 4, 3, 2 has mode 7.

Example 6

7, 3, 2 has no mode.

Example 7

7, 7, 3, 2, 2 has two modes: 7 and 2. We say that these data are *bimodal*.

?   What do you suppose data with three modes are called?

Midrange

The **midrange** is the average of the highest and lowest data.

✦   Midrange = (highest score + lowest score)/2

Example 8

From the data in Example 7, the midrange is (7 + 2) /2 = 4.5

Weighted Mean

✦   Weighted mean $= \dfrac{\Sigma\, wx}{\Sigma\, w}$, where $x_1, x_2, ..., x_n$ are data with corresponding weights

$w_1, w_2, ..., w_n$.

?   Which other measure of central tendency did we study that is basically the same as the weighted mean?

Example 9

For a certain climate comfort index, sunny days are worth three times as much as rainy days. Overcast days are worth twice as much as rainy days. What is the comfort index reading for a city with 45 sunny days, 23 overcast days, and 12 rainy days?

$$\text{Weighted mean } = \frac{\Sigma\, wx}{\Sigma\, w} = \frac{3\cdot45+2\cdot23+1\cdot12}{3+2+1} = \frac{135+46+12}{6} = 32.17$$

Exercises

1. Compute (a) mean, (b) median, (c) mode, and (d) midrange of these data:
    29, 20, 29, 21, 15, 10, 11, 24, 15, 16

2. Compute the mean of these data:

Ounces	Frequency
1-2	29
3-4	51
5-6	59
7-8	45
9-10	16

3. Suppose a statistics test makes up 50% of a student's course grade and two quizzes each make up 25% of the grade. If the student gets 85 points on the test and 87 and 78 points on the two quizzes, find the student's course grade.

Dispersion

Dispersion Statistics

Whereas a measure of central tendency (average) gives a value that tends to be the center score, a measure of dispersion (variation) shows the *extent to which data are spread around the average value*. We will study *three* measures of dispersion.

Range

The **range** is the difference between the lowest and the highest of the observations.

Example 1

The range for 30, 50, 60, 70, 90, 100 is $100 - 30 = 70$

Example 2

The range for 25, 47, 51, 54, 95 is $95 - 25 = 70$

? Based on the two previous examples, does range seem to be a good indicator of dispersion? Why or why not?

Variance for Ungrouped Data

The **variance for *ungrouped* data** is:

◆
$$s^2 = \frac{\Sigma(x - \bar{x})^2}{n-1} = \frac{n(\Sigma x^2) - (\Sigma x)^2}{n(n-1)}$$

! The last expression above does *not* need the mean, $\bar{x}$. This comes in handy when the mean is not available.

Example 3

Find the variance of 3, 4, 7, 10.

x	$\bar{x}$	$(x-\bar{x})$	$(x-\bar{x})^2$
3	6	-3	9
4	6	-2	4
7	6	1	1
10	6	4	16
		0	30

So, $s^2 = \dfrac{\Sigma(x-\bar{x})^2}{n-1} = \dfrac{30}{3} = 10$

✔ Total of this column, called **deviations**, always equals 0.

Example 4

Find the variance of these ungrouped data.

x	$\bar{x}$	$(x-\bar{x})$	$(x-\bar{x})^2$
4	5	−1	1
4	5	−1	1
5	5	0	0
7	5	2	4
		0	6

So, $s^2 = \dfrac{\Sigma(x-\bar{x})^2}{n-1} = \dfrac{6}{3} = 2$

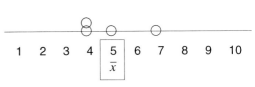

✔ Total of deviations always equals 0.

Variance for Grouped Data

For *grouped* data, $s^2 = \dfrac{\Sigma fx^2 - n\bar{x}^2}{n-1} = \dfrac{n\Sigma(fx^2)-(\Sigma fx)^2}{n(n-1)}$

◆

where x = class mark, f = class frequency, n = sample size.

! The last expression above does *not* need the mean, $\bar{x}$. This comes in handy when the mean is not available.

Example 5

Find the variance of these grouped data.

mpg	Frequency f	Class Marks x	$f{\cdot}x$	x^2	$f{\cdot}x^2$
14 to 16	9	15	135	225	2,025
16 to 18	13	17	221	289	3,757
18 to 20	24	19	456	361	8,664
20 to 22	38	21	798	441	16,758
22 to 24	16	23	368	529	8,464
✔ $\Sigma f = n = 100$			$\Sigma fx = 1{,}978$		$\Sigma fx^2 = 39{,}668$

First find the mean:

$$\bar{x} = \frac{\Sigma fx}{n} = \frac{1,978}{100} = 19.78$$

Then,

$$s^2 = \frac{39,668 - 100(19.78)^2}{100-1} = \frac{543.16}{99} = 5.486$$

or, using the alternate formula,

$$s^2 = \frac{100(39,668) - (1,978)^2}{100(100-1)} = \frac{54,316}{9,900} = 5.486$$

?   What are the first two columns in the table above called?

Standard Deviation

✦   The **standard deviation** is the positive square root of variance. That is, $s = \sqrt{s^2}$

Example 6

From Example 5, $s^2 = 5.486$. So, $s = \sqrt{5.486} = 2.34$ mpg.

?   Which of the measures of dispersion use the mean in their calculation?

!   We use Greek letters when dealing with a population and English letters with a sample:

	Mean	Variance	Standard Deviation
Sample	$\bar{x}$	s^2	s
Population	μ	σ^2	σ

Exercises

1.   Compute (a) variance, (b) standard deviation, and (c) range of these data:
     29, 20, 29, 21, 15, 10, 11, 24, 15, 16

2.   Compute (a) variance and (b) standard deviation of these data:

Ounces	Frequency
1-2	29
3-4	51
5-6	59
7-8	45
9-10	16

Position

Standard Scores (z Scores)

Standard scores, also called *z* **scores**, are used to compare data from different samples.

◆

$$z = \frac{x - \bar{x}}{s}$$

where x is the value from the sample under study. $\bar{x}$ and s are from the same sample.

! You will see the *z* score formula in one form or another throughout your study of statistics.

Example 1

Scores of three students, each from a different section of statistics, are given below. Which score is the best after standardization?

	Test 1	Test 2	Test 3
x	82	81	85
$\bar{x}$	75.25	82.50	81.75
s	9.84	3.21	5.53

$$z_1 = \frac{82 - 75.25}{9.84} = 0.69$$

$$z_2 = \frac{81 - 82.50}{3.21} = -0.47$$

$$z_3 = \frac{85 - 81.75}{5.53} = 0.59$$

Since -0.47 < 0.59 < 0.69, the raw score of $x = 82$ (corresponding to 0.69) is the best score after standardization.

Quartiles, Deciles, and Percentiles

These notions are expansions upon the role of the median.

Median divides the sample scores into 2 equal parts.

Quartiles, Q_1, Q_2, Q_3, divide the sample scores into 4 equal parts.

Deciles, D_1, D_2, ..., D_9, divide the sample scores into 10 equal parts.

Percentiles, P_1, P_2, ..., P_{99}, divide the sample scores into 100 equal parts.

! Notice how quartiles, deciles, and percentiles relate.

$$
\begin{array}{rcll}
 & & D_1 & = & P_{10} \\
 & & D_2 & = & P_{20} \\
Q_1 & = & & & P_{25} \\
 & & D_3 & = & P_{30} \\
 & & D_4 & = & P_{40} \\
\text{Median} = Q_2 & = & D_5 & = & P_{50} \\
 & & D_6 & = & P_{60} \\
 & & D_7 & = & P_{70} \\
Q_3 & = & & & P_{75} \\
 & & D_8 & = & P_{80} \\
 & & D_9 & = & P_{90} \\
\end{array}
$$

Problem Type: Given the Value, Find Its Position

◆ Given a value x,

quartile of x = (# of data less than x / total number of data) · 4, rounded off

decile of x = (# of data less than x / total number of data) · 10, rounded off

percentile of x = (# of data less than x / total number of data) · 100, rounded off

Example 2

What percentile is the score of 65 in the data shown?

Student #	Score
1	46
2	55
3	58
4	65
5	70
6	74
7	78
8	82
9	88
10	97

Three out of ten scores are less than 65. So, 65 is $\left(\dfrac{3}{10}\right) \cdot 100 = 30$th percentile $= P_{30}$.

Problem Type: Given the Position, Find the Value

Given the position either as a quartile Q_k, decile D_k, or percentile P_k, find its value by using this equation:

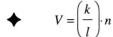

$$V = \left(\frac{k}{l}\right) \cdot n$$

where $l = 4$ for a quartile, 10 for a decile, or 100 for a percentile and n = number of data.

Arrange the data in ascending order. If V is an integer, the value of Q_k, D_k, or P_k is the mean of the Vth datum and V+1st datum.

If V is not an integer, *round up* V. Then, the value of Q_k, D_k, or P_k is the Vth datum.

Example 3

What is the value of P_{45} in the data in Example 2 above?

$$V = \left(\frac{k}{l}\right) \cdot n = \left(\frac{45}{100}\right) \cdot 10 = 4.5$$

Round *up* to 5. The 5th score is 70. So, the value of P_{45} is 70.

Exercises

1. Weights of 16-year-old boys have a mean of 145.3 lbs. and a standard deviation of 10.2 lbs. Find the z score of a 16-year-old boy weighing 147.0 lbs.

For exercises 2 and 3, use the following data: 33, 38, 43, 42, 35, 41, 34, 37, 30, 31, 45, 41

2. For the value 35, find its (a) percentile, (b) decile, (c) quartile.

3. Find (a) P_{74}, (b) D_3, (c) Q_3

Solutions to Exercises

Central Tendency

1. a) $\bar{x} = \dfrac{\Sigma x}{n} = \dfrac{29+20+29+21+15+10+11+24+15+16}{10} = 19$

Arrange the data in order of magnitude for parts (b) through (d): 10, 11, 15, 15, 16, 20, 21, 24, 29, 29.

b) Since there are an even number of data, the median is the mean of the two middle values: $\dfrac{16+20}{2} = 18$

c) 15 and 29 are both modes since they occur most often, twice each.

d) $\dfrac{10+29}{2} = 19.5$

2. $\bar{x} = \dfrac{\Sigma fx}{n} = \dfrac{29(1.5)+51(3.5)+59(5.5)+45(7.5)+16(9.5)}{29+51+59+45+16} = \dfrac{1036}{200} = 5.18$

3. $\dfrac{\Sigma wx}{\Sigma w} = \dfrac{2(85)+1(87)+1(78)}{2+1+1} = \dfrac{335}{4} = 83.75$

Dispersion

1. a) $s^2 = \dfrac{\Sigma(x-\bar{x})^2}{n-1} = \dfrac{(29-19)^2+(20-19)^2+...+(15-19)^2+(16-19)^2}{10-1} = \dfrac{416}{9} = 46.22$

   b) $s = \sqrt{46.22} = 6.80$

   c) $29 - 10 = 19$

2. a) $s^2 = \dfrac{\Sigma fx^2 - n\bar{x}^2}{n-1} = \dfrac{6450-200(5.18^2)}{200-1} = \dfrac{6450-5366.48}{199} = 5.44$

   b) $s = \sqrt{5.44} = 2.33$

Position

1. $z = \dfrac{147 - 145.3}{10.2} = 0.167$

For exercises 2 and 3, arrange the data in order of magnitude:
30, 31, 33, 34, 35, 37, 38, 41, 41, 42, 43, 45

2. a) $\dfrac{4}{12} \cdot 100 = 33.\overline{3} = P_{33}$

   b) $\dfrac{4}{12} \cdot 10 = 3.\overline{3} = D_3$

   c) $\dfrac{4}{12} \cdot 4 = 1.\overline{3} = Q_1$

3. a) $\dfrac{74}{100} \cdot 12 = 8.88 \approx 9$. So, $P_{74} = 41$

   b) $\dfrac{3}{10} \cdot 12 = 3.6 \approx 4$. So, $D_3 = 34$

   c) $\dfrac{3}{4} \cdot 12 = 9$. So, $Q_3 = 41$

Answers to ?

Page	Answer
3.1	see Grouped and Ungrouped Data, page 2.1
3.1	see Basic Terms, page 1.3
3.2	trimodal
3.3	mean for grouped data
3.4	No, since it considers only the first and last data, disregarding the middle data.
3.5	Since the differences are based on the mean (the central value), negative and positive deviations from the other data must cancel each other.
3.5	No, variance is indicated by how much the data points are spread around the mean. Variance is relative to the particular graph.
3.6	frequency distribution
3.6	variance and standard deviation

4. Probability and Counting

Probability Basics

Experiment

An **experiment** is a process which results in **outcomes (observations)**. The collection of all possible outcomes is called a **sample space**. We can perform the experiment several times. These repetitions are called **trials**.

Following are examples of experiments and their trials and outcomes.

Example 1

A coin is tossed. The sample space is {H, T}, where H = heads and T = tails. If we toss the coin again, we have performed two trials of the experiment.

Example 2

Two dice are rolled. The sample space for this single trial of the experiment is:

1,1	1,2	1,3	1,4	1,5	1,6
2,1	2,2	2,3	2,4	2,5	2,6
3,1	3,2	3,3	3,4	3,5	3,6
4,1	4,2	4,3	4,4	4,5	4,6
5,1	5,2	5,3	5,4	5,5	5,6
6,1	6,2	6,3	6,4	6,5	6,6

Event

An **event** is any subset of a sample space.

Probability of an Event

◆ If m out of the n *equally likely* outcomes of an experiment pertain to an event A, then

the **probability of A** is $P(A) = \dfrac{m}{n}$.

Example 3

From Example 1, the probability of tossing two heads is $P(\text{HH}) = \dfrac{1}{4}$.

Example 4

From Example 2, the probability of rolling a total of 7 is $P(\text{dice total 7}) = \dfrac{6}{36} = \dfrac{1}{6}$.

Notes

! For any event A, $P(A)$ has a value between 0.00 and 1.00, inclusively. That is, $0 \leq P(A) \leq 1$

! If all outcomes in the sample space pertain to event A, then $P(A) = n/n = 1$ and A is certain to happen. That is, $P(\text{not } A) = 0$.

Using Trees to List Outcomes

Trees graphically show all possible outcomes of an experiment. We define a tree through the next example.

Example 5

List all possible outcomes for the sexes of three children.

First child	Second child	Third child	Outcome
	M	M	MMM
		F	MMF
M	F	M	MFM
		F	MFF
F	M	M	FMM
		F	FMF
	F	M	FFM
		F	FFF

Finding the Number of Outcomes

It is sometimes possible to view an experiment as a series of trials of a simple experiment, where the outcomes of the trials do not affect one another. In this case, the **number of outcomes** of the overall experiment is:

◆
$$\text{number of outcomes} = a^b$$
where a is the number of outcomes per trial, and
b is the number of trials in the overall experiment.

✔ Use this formula to ensure that you've listed all possible outcomes for your experiment.

Example 6

In Example 5, the overall experiment of determining the sexes of three children can be viewed as three trials of the simple experiment of determining one child's sex. The simple experiment has a sample space of {M, F}. Repeated trials of the simple experiment do not change this sample space. Using the formula above, the number of outcomes of the overall experiment is $2^3 = 8$.

How many possible outcomes are there for the sexes of four children?

The Complement of an Event

The **complement of an event** is the event that A will not occur. The complement of event A is written A^c. A and A^c are mutually exclusive.

!   That is, $P(A) + P(A^c) = 1$. So, $P(A) = 1 - P(A^c)$ and $P(A^c) = 1 - P(A)$.

The **Venn diagram** to the right shows this relationship.

Example 7

Use a list of outcomes in two ways to find the probability of getting at least one head in three tosses of a coin.

By using a tree, we find the outcomes to be HHH, HHT, HTH, HTT, THH, THT, TTH, TTT.

So, $P(A) = P(\text{at least one H}) = \dfrac{7}{8}$.

Using the complementary event, $P(A^c) = P(\text{no H}) = P(\text{all T's}) = 1 - \dfrac{7}{8} = \dfrac{1}{8}$.

Exercises

Use the following for exercises 1 through 3:
Each of three movie reviewers rates films as P for poor, A for average, or G for good. Then they combine their individual ratings into a composite rating, such as AAG.

1. How many composite ratings (outcomes) can there be for a single movie rated by all three reviewers? List these outcomes.

2. What is the probability that the reviewers give the movie exactly two G ratings?

3. What is the probability that the reviewers give the movie at least two P ratings?

4. Use a complementary event to find the probability that the reviewers give the movie at least one P rating or at least one G rating.

Addition Rule

Symbols

 The symbol $\cup$ is equivalent to the concept of *union* of sets, the operation of *addition*, and the word *"or."* Notice how these ideas are used in this section.

Disjoint Events

Events *A* and *B* are **disjoint** if they cannot occur simultaneously.

Addition Rule for Disjoint Events

The **Addition Rule** for disjoint events A and B is:

◆ $$P(A \text{ or } B) = P(A \cup B) = P(A) + P(B)$$

Example 1

In a standard card deck (52 cards), find the probability of drawing an ace *or* a face card. These events are disjoint. That is, they cannot occur simultaneously.

P(drawing an ace *or* a face card) = 4/52 + 12/52 = 1/13 + 3/13 = 4/13 = 0.308 = 30.8%

Addition Rule for Non-Disjoint Events

If events *A* and *B* are *not* disjoint, the **Addition Rule** is:

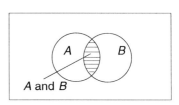

◆ $$P(A \text{ or } B) = P(A \cup B) = P(A) + P(B) - P(A \text{ and } B)$$

Example 2

In a standard card deck, find the probability of drawing an ace *or* a heart. These events are not disjoint. That is, they can occur simultaneously.

P(drawing an ace *or* drawing a heart) = 4/52 + 13/52 – 1/52 = 16/52 = 4/13 = 30.8%

Exercises

1. A fair die is rolled once. Find the probability of rolling a 2 or a 3.

2. A fair die is rolled once. Find the probability of rolling a 2 or an even number.

Multiplication Rule

Symbols

! The symbol ∩ is equivalent to the concept of *intersection* of sets, the operation of *multiplication,* and the word *"and"*. Notice how these ideas are used in this section.

Independent Events

If the occurrence of event A does not affect the probability that event B will occur, then A and B are said to be **independent**.

Multiplication Rule for Independent Events

The **multiplication rule for independent events** A and B is:

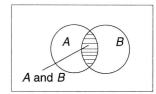

♦ $$P(A \text{ and } B) = P(A \cap B) = P(A) \cdot P(B)$$

Example 1

Two cards are drawn from a standard deck. Before the second card is drawn, the first card is replaced. In this way, the second draw is not affected by one less card in the deck from the first draw. That is, the events are independent. Find the probability of drawing two kings.

$P(\text{drawing two kings}) = P(\text{king on first } and \text{ second draws})$

$$= \frac{4}{52} \cdot \frac{4}{52} = \frac{1}{13} \cdot \frac{1}{13} = \frac{1}{169} = 0.006 = 0.6\%$$

Example 2

A die is thrown three times. Find the probability of throwing a 2 on all three throws.

$P(\text{throwing 2 on all three throws}) = P(2 \text{ on first } and \text{ second } and \text{ third throws})$

$$= \frac{1}{6} \cdot \frac{1}{6} \cdot \frac{1}{6} = \frac{1}{216} = 0.005 = 0.5\%$$

Conditional Probability

The probability of event B occurring given that event A has occurred, $P(B|A)$, is called **conditional probability**.

! $B|A$ does not indicate division.

Multiplication Rule with Conditional Probability

When the probability of event B depends on event A occurring, the **multiplication rule** is:

✦ $P(A \text{ and } B) = P(A \cap B) = P(A) \cdot P(B|A)$

Example 3

Two cards are drawn from a deck *without replacement.* So, the second draw is affected by one less card in the deck from the first draw. So, the second draw is conditional upon the first draw. Find the probability of drawing two kings.

$P(\text{drawing two kings}) = P(\text{king on the first } and \text{ second draws})$

$$= \frac{4}{52} \cdot \frac{3}{51} = \frac{1}{221} = 0.005 = 0.5\%$$

Example 4

A box contains three white balls and two black balls. Two balls are selected *without replacement.*

$P(\text{first ball is black } and \text{ the second ball is also black}) = \frac{2}{5} \cdot \frac{1}{4} = \frac{1}{10} = 0.1 = 10\%$

Multiplication Rule and Complementary Events

Some multiplication rule problems can be solved by using the complementary event

Example 5

Nationally, 15% of all grade school children are in fourth grade. If two grade school children are randomly chosen, find the probability that at least one of them is in fourth grade.

If $P(A) = P(\text{at least one child is in fourth grade})$
then $P(A^c) = P(\text{neither is in fourth grade}) = 0.85 \cdot 0.85 = 0.72$
and $P(A) = 1 - P(A^c) = 1 - 0.72 = 0.28$

?   What is $P(A) + P(A^c)$ equal to?

Exercises

1. On average, a baseball team wins 6 out of every 10 games that it plays. What is the probability that the team wins its next three games?

2. There are two winning raffle tickets in a hat containing 10 tickets. What is the probability of drawing both winning tickets in two draws?

3. From exercise 2, use a complementary event to find the probability of drawing at least one winning ticket in two draws.

Bayes' Theorem

In some experiments, the probability of the outcome of an event A might be influenced by another event B. Then, we can "condition" the probability of event A by using Bayes' Theorem:

$$P(A|B) = \frac{P(B|A) \cdot P(A)}{P(B|A) \cdot P(A) + P(B|A^c) \cdot P(A^c)}$$

Example

A study shows that 40% of the population eats at least 15 grams of fiber daily. Of those people, 5% develop a serious intestinal problem in their lifetime. Those who do not eat at least 15 grams of fiber daily are 10% likely to develop a serious intestinal problem. Given that a randomly selected person has developed a serious intestinal problem, find the conditional probability that he or she eats at least 15 grams of fiber daily .

Tabulate the data as follows:

	Proportion of population	Proportion with intestinal problem
Eat ≥ 15 g	0.40	0.05
Eat < 15g	0.60	0.10
Total	1.00	

✔ This column always totals 1.00.

Let A = Person eats at least 15 grams of fiber.

Let B = Person has serious intestinal problem.

Then, $P(A|B) = \dfrac{0.05 \cdot 0.40}{0.05 \cdot 0.40 + 0.10 \cdot 0.60} = \dfrac{0.02}{0.02 + 0.06} = 0.25$

Exercise

In a mail order house, a clerk processes 60% of the orders. A second clerk processes 40% of the orders. The first clerk has an error rate of 1%. The second clerk has an error rate of 2%. A processed order is randomly selected and found to have an error. What is the probability that the first clerk made the error?

Methods of Counting

Fundamental Rule of Counting

The **Fundamental Rule of Counting** says that if an action can be performed in m ways and another action can be performed n ways, then both actions can be performed successfully $m \cdot n$ ways.

Example 1

Five airlines offer service from Chicago to New York and three offer service from New York to Frankfurt. How many different flight arrangements can be made from Chicago to Frankfurt?

Chicago ———————●New York ———————● Frankfurt
　　　　　5 flights　　　　　　　　　3 flights

There are 5·3=15 different flight arrangements.

Factorials

Factorials are the first tools we use to develop the formula for determining probabilities of binomial distributions. The symbol we use is $n!$, read "n factorial," where n is a whole number. In general,

◆
$$n! = n(n-1)(n-2) \cdots 2 \cdot 1$$

! There is one exception to this rule: $0! = 1$

Example 2

$$5! = 5 \cdot 4 \cdot 3 \cdot 2 \cdot 1 = 120$$

Combinations

Combinations use factorials to count how many ways items can be selected when the order in which the items are selected is *not* important.

◆
$$_nC_i = \frac{n!}{i!(n-i)!}$$

where n = total number of items, and
i = number of items pertaining to an event A.

Example 3

How many ways can a 100-member club elect a 4-person committee?

In this example, $n = 100$ and $i = 4$. So, we have

$$_{100}C_4 = \frac{100!}{4!(100-4)!} = \frac{100!}{4!96!} = \frac{100 \cdot 99 \cdot 98 \cdot 97 \cdot \cancel{96 \cdot 95 \cdots 2 \cdot 1}}{(4 \cdot 3 \cdot 2 \cdot 1)\cancel{(96 \cdot 95 \cdots 2 \cdot 1)}} = \frac{100 \cdot 99 \cdot 98 \cdot 97}{4 \cdot 3 \cdot 2 \cdot 1}$$
$$= 3{,}921{,}225 \text{ ways.}$$

! Never divide out factorials. For example, $\dfrac{10!}{2!} \neq 5!$ You must rewrite them as products first.

Permutations

Another way to count how many ways items can be selected is called **permutations**. Use permutations when the order of a selection *is* important.

◆
$$_{n}P_i = \frac{n!}{(n-i)!}$$

where n = total number of items, and
i = number of items pertaining to an event A.

   How is the formula for combinations different from that of permutations?

Example 4

How many ways can a 100 member club elect four officers? The top vote-getter becomes President; the second highest vote-getter becomes Vice President; third highest, Secretary; fourth highest, Treasurer.

In this example, $n = 100$ and $i = 4$. So, we have

$$_{100}P_4 = \frac{100!}{(100-4)!} = \frac{100!}{96!} = \frac{100 \cdot 99 \cdot 98 \cdot 97 \cdot \cancel{96 \cdot 95 \cdots 2 \cdot 1}}{\cancel{96 \cdot 95 \cdots 2 \cdot 1}} = 100 \cdot 99 \cdot 98 \cdot 97 = 94{,}109{,}400$$

   Compare the answers to Examples 3 and 4. Why are there more permutations than combinations?

Exercises

1. A sample of people are grouped according to hair and eye color. How many categories are possible if hair color can be brown, black, blonde, or red, and eye color can be brown, blue, or green?

2. How many different 5 person lineups can a basketball coach make from a roster of 12 if the players' positions is not important?

3. How many different 5 person lineups can a basketball coach make from a roster of 12 if the players' positions is important?

Solutions to Exercises

Probability Basics

List of outcomes:

1. $3^3 = 27$ 2. $P(\text{two G's}) = \dfrac{6}{27} = 0.222$

3. $P(\text{at least two P's}) = \dfrac{7}{27} = 0.259$

4. $P(A) = P(\text{at least one P or at least one G})$

$P(A^c) = P(\text{no P and no G}) = P(\text{All A's}) = \dfrac{1}{27} = 0.037$

So, $P(A) = 1 - P(A^c) = 1 - \dfrac{1}{27} = \dfrac{26}{27} = 0.963$

PPP	APP	GPP
PPA	APA	GPA
PPG	APG	GPG
PAP	AAP	GAP
PAA	AAA	GAA
PAG	AAG	GAG
PGP	AGP	GGP
PGA	AGA	GGA
PGG	AGG	GGG

Addition Rule

1. $P(\text{rolling 2 or 3}) = \dfrac{1}{6} + \dfrac{1}{6} = \dfrac{1}{3} = 0.333$

2. $P(\text{rolling 2 or even number}) = \dfrac{1}{6} + \dfrac{3}{6} - \dfrac{1}{6} = \dfrac{1}{2} = 0.500$

Multiplication Rule

1. $P(\text{win next three games}) = \dfrac{6}{10} \cdot \dfrac{6}{10} \cdot \dfrac{6}{10} = \dfrac{216}{1000} = 0.216$

2. $P(\text{both winners}) = \dfrac{2}{10} \cdot \dfrac{1}{9} = \dfrac{1}{45} = 0.022$

3. $P(\text{at least one winner in two draws}) = 1 - P(\text{No winners in two draws})$

$= 1 - \left(\dfrac{8}{10} \cdot \dfrac{7}{9} \right) = 1 - \dfrac{28}{45} = \dfrac{17}{45} = 0.378$

Bayes' Theorem

Let A = First clerk processed the order. Let B = The order has an error.

$P(A|B) = \dfrac{(0.01)(0.60)}{(0.01)(0.60) + (0.02)(0.40)} = 0.429$

Methods of Counting

1. $4 \cdot 3 = 12$

2. $\displaystyle {}_{12}C_5 = \frac{12!}{5!(12-5)!} = \frac{12}{5!7!} = \frac{12 \cdot 11 \cdot 10 \cdot 9 \cdot 8 \cdot \cancel{7 \cdot 6 \cdot 5 \cdot 4 \cdot 3 \cdot 2 \cdot 1}}{5 \cdot 4 \cdot 3 \cdot 2 \cdot 1 \cdot \cancel{7 \cdot 6 \cdot 5 \cdot 4 \cdot 3 \cdot 2 \cdot 1}} = \frac{12 \cdot 11 \cdot 10 \cdot 9 \cdot 8}{5 \cdot 4 \cdot 3 \cdot 2 \cdot 1} = 792$

3. $\displaystyle {}_{12}P_5 = \frac{12!}{(12-5)!} = \frac{12}{7!} = \frac{12 \cdot 11 \cdot 10 \cdot 9 \cdot 8 \cdot \cancel{7 \cdot 6 \cdot 5 \cdot 4 \cdot 3 \cdot 2 \cdot 1}}{\cancel{7 \cdot 6 \cdot 5 \cdot 4 \cdot 3 \cdot 2 \cdot 1}} = 12 \cdot 11 \cdot 10 \cdot 9 \cdot 8 = 95{,}040$

Answers to ?

Page	Answer
4.2	16
4.6	1.00; see The Complement of an Event, p. 4.3
4.9	The combination formula contains the $i!$ factor.
4.9	The larger number of permutations indicates that the order in which the officers are elected is important. For example, Joe, Mary, Fred, and June as President, Vice President, Secretary, and Treasurer, respectively, is different from June, Fred, Mary, and Joe as President, Vice President, Secretary, and Treasurer, respectively. In terms of combinations, there is only one committee consisting of Joe, Mary, Fred, and June regardless of the number of votes received by each member.

5. Discrete Random Variables

Probability Distributions

Random Variable

A **random variable** is a quantity that may take on several values.

! Think of a random variable as you do a variable in algebra.

A **discrete random variable** is one in which there is a "gap" or "interruption" in the values that the variable can assume. A discrete random variable has a finite number of values.

A **continuous random variable** is one which theoretically can assume any value within an interval of values. A continuous random variable has an infinite number of values.

Example 1

In a throw of a die, the number of dots coming up is a *discrete* random variable since it cannot take on values such as 1.5.

Example 2

Time is a *continuous* random variable.

? What is a relative frequency distribution?

Probability Distribution

A **probability distribution** is a relative frequency distribution for a population. It is a collection of values for a random variable, x, and the corresponding probabilities, $P(x)$. For all $P(x)$ in a probability distribution, two criteria must be met:

1) $0 \le P(x) \le 1$, for each x and   2) $\sum P(x) = 1$

? What is another name for the collection of probabilities, $P(x)$?

Example 3

When tossing a coin, $P(H) = P(T) = 1/2$. The probability distribution is:

x	$P(x)$
H	1/2
T	1/2
	2/2 = 1

? What is a histogram?

Example 4

The probability distribution for the heights of a group of ponies and its bar chart are:

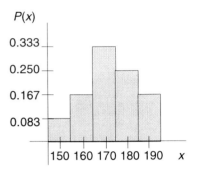

Height in cm x	Frequency f	Relative Frequency $P(x)$
150	1	0.083
160	2	0.167
170	4	0.333
180	3	0.250
190	2	0.167
	12	1.000

✔ This column always totals *n*.

✔ This column always totals 1.

Example 5

A die is rolled. The probability distribution and bar chart are shown.

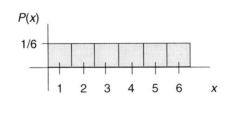

x	$P(x)$
1	1/6
2	1/6
3	1/6
4	1/6
5	1/6
6	1/6
	6/6 = 1

! A probability distribution where all $P(x)$ are equal is called a **uniform probability distribution**.

! If the width of the interval between values of x equals 1, then the area of the rectangle equals $P(x)$. Also, the sum of the areas of *all* rectangles is $\sum P(x) = 1$.

Exercises

1. Are the values of ten-digit telephone numbers discrete or continuous? Defend your answer.

Do exercises 2 and 3 represent probability distributions? Defend your answers.

2.

x	$P(x)$
2	0.33
4	0.67
6	0.33

3.

x	$P(x)$
2	25
4	50
6	25

Describing Probability Distributions

As we did with grouped and ungrouped data, we can talk about *central tendency* and *dispersion* with random variables.

? What is a weighted mean?

Expected Value

The **expected value** of a discrete random variable x is the average of all possible values of x weighted by their probabilities. We use expected value, $E(x)$, to measure central tendency of a random variable.

◆
$$E(x) = \mu = \Sigma x \cdot P(x)$$

! Expected value is the *mean* of the random variable x. It is also called the *expectation of* x. In the physical sciences, expected value is interpreted as the center of gravity.

Example 1

A coin is tossed. If the coin turns up heads, you win $10; tails, you lose $10. What is the expected value of the game?

The values of x are positive if they represent profits (winnings) and negative if they represent deficits (losses). So, in this example, x can take on the value $10 or –$10.

Since, $P(\text{H}) = P(\text{T}) = \dfrac{1}{2}$, we have $E(x) = \$10 \cdot \left(\dfrac{1}{2}\right) + (-\$10) \cdot \left(\dfrac{1}{2}\right) = 0$.

Therefore, the game is fair since no one is favored.

! When $E(x) = 0$, the game is said to be fair, or equitable.

Example 2

One thousand raffle tickets are sold for $1 each. Four prizes are awarded: $200, $150, $100, and $50. If you purchase a ticket, what is your expected gain?

x can take on the value $199, $149, $99, $49 or –$1. So,

$$P(1\text{st}) = P(2\text{nd}) = P(3\text{rd}) = P(4\text{th}) = \frac{1}{1000} \qquad P(\text{no prize}) = \frac{996}{1000}$$

$$E(x) = \$199\left(\frac{1}{1000}\right) + \$149\left(\frac{1}{1000}\right) + \$99\left(\frac{1}{1000}\right) + \$49\left(\frac{1}{1000}\right) + (-\$1)\left(\frac{996}{1000}\right) = -\$0.50$$

That is, over the long run, there is an average loss of $0.50. Thus, the game is not fair.

Variance and Standard Deviation

Variance and standard deviation for a discrete random variable x are:

$$\blacklozenge \quad Var(x) = \sigma^2 = \left[\sum x^2 \cdot P(x)\right] - \mu^2 = E(x^2) - \mu^2 \qquad SD(x) = \sigma = \sqrt{Var(x)}$$

Example 3

This is a probability distribution for the number of television sets per family as determined by a national poll. Find the expected value, variance, and standard deviation.

Number of TV Sets x	$P(x)$	$x \cdot P(x)$	x^2	$x^2 \cdot P(x)$
0	0.2	0.0	0	0.0
1	0.2	0.2	1	0.2
2	0.3	0.6	4	1.2
3	0.1	0.3	9	0.9
4	0.1	0.4	16	1.6
5	0.1	0.5	25	2.5
	1.0	$\mu = 2.0$		$E(x^2) = 6.4$

$$E(x) = \mu = \sum x \cdot P(x) = 2.0$$

$$Var(x) = \sigma^2 = E(x^2) - \mu^2 = \sum \left(x^2 \cdot P(x)\right) - \mu^2 = 6.4 - (2.0)^2 = 2.4$$

$$SD(x) = \sigma = \sqrt{2.4} = 1.6$$

Properties of Expected Value and Variance

These properties come in handy when dealing with more than one random variable. Think of these properties in terms of mean and variance for grouped and ungrouped data to help understand them.

For expected value (mean):   For variance:

$$E(cx) = cE(x) \qquad (1) \qquad\qquad Var(cx) = c^2 Var(x) \qquad (4)$$

$$E(x+c) = E(x) + c \qquad (2) \qquad\qquad Var(x+c) = Var(x) \qquad (5)$$

$$E(x+y) = E(x) + E(y) \qquad (3) \qquad\qquad Var(x+y) = Var(x) + Var(y) \qquad (6)\ *$$

where x and y are random variables, and c is a constant.

* Equation (6) holds only when x and y are independent, that is, when knowing the value of one does not influence the probability of the other.

?   In terms of probabilities, what are independent events?

Example 4

Suppose the expected value of a probability distribution for the random variable x is 4.7 and its variance is 0.2. Also suppose the expected value of a probability distribution for the random variable y is 5.3 and its variance is 1.1.

If each value of x is multiplied by 3, the new expected value is $E(3x) = 3E(x) = 3(4.7) = 14.1$ (equation (1)). Its new variance is $Var(3x) = 3^2Var(x) = 9(0.2) = 1.8$ (equation (4)).

If 3 is added to each value of x, the new expected value is $E(x + 3) = E(x) + 3 = 4.7 + 3 = 7.7$ (equation (2)). Its variance remains the same: $Var(x + 3) = Var(x) = 0.2$ (equation (5)).

Also, $E(x + y) = 4.7 + 5.3 = 10.0$ (equation (3)). $Var(x + y) = 0.2 + 1.1 = 1.3$ only if x and y are independent (equation (6)).

Exercises

During a summer in a certain region of the country, these is a 50% chance of a sunny day, a 30% chance of a cloudy day without precipitation, and a 20% chance of a cloudy day with precipitation. On a sunny day, 250 people will visit the local beach. On a cloudy day without precipitation, 100 people will visit the local beach. On a cloudy day with precipitation, 10 people will visit the local beach.

1. What is the expected number of people at the beach on a summer day?

2. Find $Var(x)$ and $SD(x)$ for this scenario.

3. Suppose all three headcounts are increased by 10. Find $E(x)$ and $Var(x)$ for these new numbers.

Binomial Distributions

We now consider a specific kind of probability distribution called a **binomial distribution**. A binomial distribution involves a *discrete* random variable.

Binomial Experiment

A **binomial experiment** is one in which there are just two outcomes: some event A either happens or does not happen. The corresponding distribution is called a binomial distribution.

Example 1

Experiments with these outcomes are binomial: heads or tails, win or lose, good or defective.

Counterexample 1

A hockey game is *not* a binomial experiment since it can result in a win, loss, or tie. Throwing a die is *not* a binomial experiment if we consider that there are six possible outcomes.

Binomial Probability Formula

If an experiment is performed n independent times, each trial having two and only two outcomes, say, A and not A, and if $P(A) = p$ and $P(\text{not } A) = q = 1 - p$, then

✦
$$P(A \text{ occurs } i \text{ times}) = {}_nC_i \cdot p^i \cdot q^{n-i}$$
where n = number of trials
        i = number of successes in n trials
        $p = P(A)$ = probability of success
        $q = 1 - p$ = probability of failure

This is the **binomial probability formula**.

❗ A good first step in solving a binomial probability is to identify the values of n, i, p, and q.

Example 2

A quiz has six multiple-choice questions, each with three choices. If you guess at all questions, find the probability of answering all questions correctly.

There are $n = 6$ questions We want $i = 6$ answered correctly. The probability of guessing a correct answer is $p = \dfrac{1}{3} = 0.33$, and $q = 1 - p = 1 - 0.33 = 0.67$. So,

$P(\text{all correct}) = {}_6C_6(0.33)^6(0.67)^{6-6} = 1(0.33)^6(0.67)^0 = 0.001 = 0.1\%$

Example 3

Two coins are tossed simultaneously five times. Find the probability that double heads will appear two times.

Out of $n = 5$ tosses, we want $i = 2$ successes. In each simultaneous toss of both coins, $P(\text{HH}) = p = 0.25$. So, $q = 1 - p = 1 - 0.25 = 0.75$.

Thus, $P(2\ \text{HH}) = {}_5C_2(0.25)^2(0.75)^{5-2} = 10(0.25)^2(0.75)^3 = 0.264$

Example 4

Two coins are tossed simultaneously five times. Find the probability that double heads will appear *three or more* times.

$$P(\text{HH three or more times}) = P(3\ \text{HH}) + P(4\ \text{HH}) + P(5\ \text{HH})$$

$$= {}_5C_3(0.25)^3(0.75)^{5-3} + {}_5C_4(0.25)^4(0.75)^{5-4} + {}_5C_5(0.25)^5(0.75)^{5-5}$$

$$= 10(0.25)^3(0.75)^2 + 5(0.25)^4(0.75)^1 + 1(0.25)^5(0.75)^0$$

$$= 0.088 + 0.015 + 0.001 = 0.104 = 10.4\%$$

Example 5

Suppose that annually seven out of ten days are sunny in a certain region of the country. What is the probability that three randomly selected days are sunny?

$P(\text{all three days are sunny}) = {}_3C_3(0.7)^3(0.3)^0 = 0.343$

What is the probability that none of the three days are sunny?

$P(\text{no days are sunny}) = {}_3C_0(0.7)^0(0.3)^3 = 0.027$

What is the probability that at least one day of the three is sunny?

$$P(\text{at least one day is sunny}) = {}_3C_1(0.7)^1(0.3)^2 + {}_3C_2(0.7)^2(0.3)^1 + {}_3C_3(0.7)^3(0.3)^0$$

$$= (3)(0.7)(0.09) + (3)(0.49)(0.3) + (1)(0.343)(1) = 0.973$$

or, alternatively,
$P(\text{at least one day is sunny}) = 1 - P(\text{no days are sunny}) = 1 - 0.027 = 0.973$

?   What type of probability event is used in the line above?

! Use the **binomial distribution probabilities table** in the back of your text to save time solving *some* of the problems in this section. This table might not be useful for all problems since it might not contain all values for n, i, p, and q.

Exercises

Answer exercises 1-3 based on this scenario: A baseball team has an 0.600 probability of winning when challenging their crosstown rivals. Next weekend, the team begins a 4-game series against the crosstown rivals.

1. What is the probability that the team wins exactly 3 games in this series?

2. What is the probability that the team wins at least 3 games in this series?

3. What is the probability that the team wins at least 1 game in this series? Solve this exercise with the binomial theorem, then with a complementary event. Your answers for both methods should be the same.

Describing Binomial Distributions

Parameters of a Binomial Distribution

? What is a parameter?

Expected value, variance, and standard deviation of a binomial distribution could be calculated by using the general formulas for a probability distribution, $E(x) = \sum x \cdot P(x)$ and $Var(x) = \left[\sum x^2 \cdot P(x)\right] - \mu^2$. However, we have these specific formulas for a binomial distribution:

$$E(x) = np \quad \text{and} \quad Var(x) = npq$$

As before, $SD(x) = \sqrt{Var(x)}$

where n = number of trials
  p = favorable probability
  q = unfavorable probability

Example 1

Find the expected value, variance, and standard deviation of the random variable x representing the number of heads in four coin tosses.

$$E(x) = np = 4 \cdot \left(\frac{1}{2}\right) = 2, \; Var(x) = npq = 4 \cdot \left(\frac{1}{2}\right)\left(\frac{1}{2}\right) = 1, \text{ and } SD(x) = \sqrt{1} = 1$$

Using the general formulas for a probability distribution, we get the same answers:

$$E(x) = \sum x \cdot P(x) = 2 \text{ and } Var(x) = \left[\sum x^2 \cdot P(x)\right] - \mu^2 = 5 - 2^2 = 1$$

Example 2

A quiz has five multiple-choice questions, each with four choices. Find the expected value and variance of guessing on all five questions.

This is a binomial distribution with $n = 5$, $p = \frac{1}{4}$, $q = \frac{3}{4}$. So, $E(x) = 5 \cdot \frac{1}{4}$ and

$$Var(x) = 5 \cdot \frac{1}{4} \cdot \frac{3}{4} = \frac{15}{16} = 0.938.$$

Exercise

A baseball team has an 0.600 probability of winning when challenging their crosstown rivals. Next weekend, the team begins a 4-game series against the crosstown rivals. Find the expected number of wins, the variance, and the standard deviation for this series.

Solutions to Exercises

Probability Distributions

1. Discrete, since there are a finite number of telephone numbers that can be derived from the digits 0 through 9.

2. No, since $\sum P(x) \neq 1$.

3. No, since $\sum P(x) \neq 1$, and for each x, $P(x)$ is not between 0 and 1, inclusively.

Describing Probability Distributions

x	$P(x)$	x^2	$x^2 \cdot P(x)$
250	0.50	62,500	31,250
100	0.30	10,000	3,000
10	0.20	100	20
			34,270

1. $E(x) = 250(0.50) + 100(0.30) + 10(0.20) = 125 + 30 + 2 = 157$

2. $Var(x) = 34{,}270 - 157^2 = 34{,}270 - 24{,}649 = 9621$. So, $SD(x) = \sqrt{9621} = 98.09$

3. $E(x + 10) = E(x) + 10 = 157 + 10 = 167$
    $Var(x + 10) = Var(x) = 9621$

Binomial Distributions

1. $n = 4$, $i = 3$, $p = 0.600$, $q = 0.400$. So, $P(x = 3 \text{ wins}) = {}_4C_3(0.600)^3(0.400)^1 = 0.3456$

2. $n = 4$, $i = 3$ or 4, $p = 0.600$, $q = 0.400$. So,
    $P(x \geq 3 \text{ wins}) = {}_4C_3(0.600)^3(0.400)^1 + {}_4C_4(0.600)^4(0.400)^0 = 0.3456 + 0.1296 = 0.4752$

3. Binomial probability formula: $n = 4$, $i = 1, 2, 3,$ or 4, $p = 0.600$, $q = 0.400$. So,
    $P(x \geq 1 \text{ win}) = {}_4C_1(0.600)^1(0.400)^3 + {}_4C_2(0.600)^2(0.400)^2 + {}_4C_3(0.600)^3(0.400)^1 +$
    ${}_4C_4(0.600)^4(0.400)^0 = 0.1536 + 0.3456 + 0.3456 + 0.1296 = 0.9744$
    Complementary event: $P(\text{win at least } 1) = 1 - P(\text{lose all } 4) = 1 - 0.400^4 = 0.9744$

Describing Binomial Distributions

$E(x) = np = (4)(0.600) = 2.4$, $Var(x) = npq = (4)(0.600)(0.400) = 0.96$, $SD(x) = 0.98$.

Answers to ?

Page	Answer
5.1	see Types of Distributions, page 2.2
5.1	relative frequency distribution
5.1	see Histograms and Relative Histograms, page 2.6
5.3	see Weighted Mean, page 3.3
5.4	see Independent Events, page 4.5
5.8	complementary event
5.9	A parameter is a number describing a characteristic of a population.

6. Normal Random Variables

Normal Probability Distributions

? What is a continuous random variable?

We now consider another specific probability distribution called a **normal distribution**. A normal distribution involves a continuous random variable.

Each continuous random variable has a curve associated with it. This curve is called the **probability density function**. The curve associated with the **normal random variable** is the familiar "bell-shaped" curve, called the **normal curve**. This normal curve is shaped by two parameters, namely the mean (expected value), μ, and the standard deviation, σ.

μ appears at the center of the normal curve. The distance between two adjacent vertical lines as shown below is σ. The normal curve is **symmetrical** about the mean. That is, the right and left sides of the curve are mirror images of each other. The total area under the curve is equal to 1. Thus, 0.5000 of the random variable values lie under the right side of the curve and 0.5000 of the values lie under the left side of the curve.

Approximation Rule

In the figure below, the numbers appearing in each standard deviation region represent the probability that a random variable value lies in that region. For example, there is a probability of 0.3413 that a value lies between the mean and one standard deviation to the right of the mean. Also, there is a probability of 0.6826 that a value lies within one standard deviation on either side of the mean. The **approximation rule** allows us to determine these probabilities using the values shown on the curve.

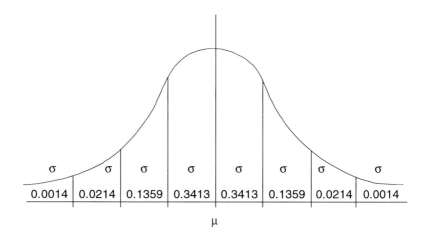

Exercise

What is the probability that a random variable value will be more than two standard deviations to the right of the mean on the normal curve shown above?

Standard Normal Probability Distributions

When a normal distribution has $\mu = 0$ and $\sigma = 1$, it is called a **standard normal distribution**.

A z score represents the number of standard deviations a value is from the mean on the **standard normal curve**.

?   What is the probability that a z score lies on the right side of a normal curve?

! Use the **standard normal distribution probabilities table** in your textbook for exercises in this chapter. *Learn how to correctly read this table.* Not all standard normal probabilities tables are alike. Some list probabilities from 0 to 0.5000, while others show 0.5000 to 1.0000. In either case, the probabilities in the tables indicate the area under the curve to the *left* of the z score. Still, your table might have a completely different structure.

Finding Standard Normal Probabilities

It is helpful, especially when first learning about the standard normal curve, to follow these steps for solving a problem involving standard normal probabilities.

Step	Action
1	Sketch the normal curve.
2	Label the curve with $\mu = 0$ and with values from your problem.
3	Shade the region(s) under the curve where the probability is to be determined.

! This visualization helps you set up the probability equation.

?   Between what two values do all probabilities lie?

Example 1

Find the probability that a z score is between 0 and 1.30.

Shade the area under the curve from 0 (the standard mean) to 1.30.

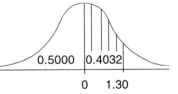

0.5000 0.4032

0   1.30

If your standard normal probabilities table lists values from:	then:
0.5000 to 1.0000	$P(0 < z < 1.30) = 0.9032 - 0.5000 = 0.4032$
0.0000 to 0.5000	$P(0 < z < 1.30) = 0.4032$

✔ probability
between 0 and 1

Example 2

Find the probability that the z score is less than 1.30.

Shade the area under the curve to the left of 1.30.

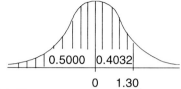

If your standard normal probabilities table lists values from:	then:
0.5000 to 1.0000	$P(z < 1.30) = 0.9032$
0.0000 to 0.5000	$P(z < 1.30) = 0.5000 + 0.4032 = 0.9032$

Example 3

Find the probability that the z score is between 0.84 and 1.30.

Shade the area under the curve from 0.84 to 1.30.

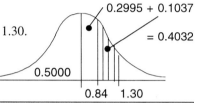

If your standard normal probabilities table lists values from:	then:
0.5000 to 1.0000	$P(0.84 < z < 1.30) = 0.9032 - 0.7995 = 0.1037$
0.0000 to 0.5000	$P(0.84 < z < 1.30) = 0.4032 - 0.2995 = 0.1037$

Example 4

Find the probability that the z score is greater than 1.30.

Shade the area under the curve to the right of 1.30.

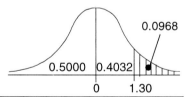

If your standard normal probabilities table lists values from:	then:
0.5000 to 1.0000	$P(z > 1.30) = 1.000 - 0.9032 = 0.0968$
0.0000 to 0.5000	$P(z > 1.30) = 0.5000 - 0.4032 = 0.0968$

Example 5

Find the probability that the absolute value of the z score is greater than 1.30.

Because of the symmetry of the normal curve,

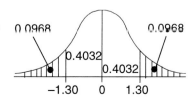

$P(|z| > 1.30) = P(z < -1.30 \text{ or } z > 1.30) = P(z < -1.30) + P(z > 1.30)$. So, shade the areas under the curve to the left of -1.30 and to the right of 1.30.

If your standard normal probabilities table lists values from:	then:
0.5000 to 1.0000	$P(|z| > 1.30)$ $= (1.0000 - 0.9032) + (1.0000 - 0.9032)$ $= 0.1936$
0.0000 to 0.5000	$P(|z| > 1.30)$ $= (0.5000 - 0.4032) + (0.5000 - 0.4032)$ $= 0.1936$

Example 6

Find the probability that the absolute value of the z score is less than 1.30.

Because of the symmetry of the normal curve, $P(|z| < 1.30) = P(-1.30 < z < 1.30)$. So, shade the region under the curve between -1.30 and 1.30.

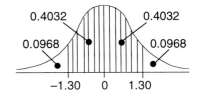

If your standard normal probabilities table lists values from:	then:
0.5000 to 1.0000	$P(|z| < 1.30)$ $= (0.9032 - 0.5000) + (0.9032 - 0.5000)$ $= 0.8064$
0.0000 to 0.5000	$P(|z| < 1.30)$ $= 0.4032 + 0.4032$ $= 0.8064$

Exercises

1. What is the probability that a z score is less than 2.16?

2. What is the probability that a z score is greater than 2.16?

3. What is the probability that a z score is between -1.34 and 2.16?

4. What is the probability that a z score is less than -1.85?

5. What is $P(|z| < 1.85)$?

Finding Normal Probabilities

To find normal probabilities, we must first convert the normal score to a z score (standard normal score). Then, using a standard normal distribution probabilities table, we find the probability that the event involving the z score will occur.

Standard Scores

In terms of a normal distribution, the standard normal score (z score) formula is:

✦
$$z = \frac{x - \mu}{\sigma}$$

? How is this formula different from the one we previously studied? What was our original use of z scores?

! We will use several forms of this basic formula throughout our study of statistics.

Example 1

The monthly expenses for college students' social lives are believed to be normally distributed with $\mu = \$100$ and $\sigma = \$15$. If a student is randomly selected, what is the probability that he will have monthly expenses greater than $136?

$$P(x > 136) = P\left(\frac{x - \mu}{\sigma} > \frac{136 - 100}{15}\right) = P(z > 2.4) = 0.0082$$

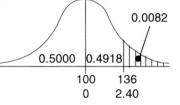

Example 2

What is the probability that the college student will have expenses less than 73?

$$P(x < 73) = P\left(\frac{x - \mu}{\sigma} < \frac{73 - 100}{15}\right) = P(z < -1.8) = 0.0359$$

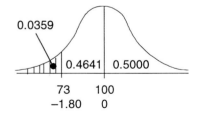

Example 3

What is the probability that the college student will have expenses between 85 and 136?

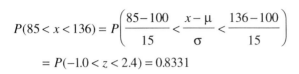

$$P(85 < x < 136) = P\left(\frac{85-100}{15} < \frac{x-\mu}{\sigma} < \frac{136-100}{15}\right)$$

$$= P(-1.0 < z < 2.4) = 0.8331$$

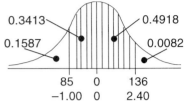

Additive Property of Normal Random Variables

Suppose x and y are independent normal random variables with parameters μ_x, σ_x, μ_y, and σ_y, respectively. Then the random variable $x + y$ is normal with

$$E(x + y) = E(x) + E(y) = \mu_x + \mu_y$$

and $$SD(x + y) = \sqrt{\sigma_x^2 + \sigma_y^2}$$

Example 4

Suppose one student is randomly selected from each of two populations of elementary school children. One population has a mean IQ of 95 and a standard deviation of 15. The other population has a mean IQ of 105 and a standard deviation of 15. What is the probability that the two children's total IQ is greater than 220?

First, $E(x + y) = 95 + 105 = 200$ and $SD(x) = \sqrt{15^2 + 15^2} = 21.21$

Then, $P(x + y > 220) = P\left(\frac{(x+y)-E(x+y)}{SD(x+y)} > \frac{220-200}{21.21}\right) = P(z > 0.94) = 0.1736$

Exercises

The weights of bags containing fresh frozen fruit have a mean of 16.1 oz and a standard deviation of 0.23 oz.

1. Find the probability that a randomly selected bag of fruit weighs at least 16.3 oz.

2. Find the probability that a randomly selected bag of fruit weighs less than 16.0 oz.

3. Find the probability that a randomly selected bag of fruit weighs between 15.9 oz and 16.2 oz.

4. If two bags of fruit are randomly selected, find the probability that they together weigh less than 32.5 oz.

Percentiles of Normal Random Variables

This section is the opposite of the previous section in that here we start with a probability and find the score. In the previous section, we started with a score and found its probability.

Percentile of a Standard Normal Distribution

For any value α between 0 and 1, z_α is the z score at the $100(1 - \alpha)$ **percentile of a standard normal distribution**.

! It is helpful, especially when first learning about the normal curve, to sketch the curve, label it with values from your problem, and shade the region under the curve where the z score is to be determined.

! We will use the value α when we construct confidence intervals and perform hypothesis testing later in this study guide.

? What percentage of scores lie to the left of the mean in a normal distribution?

Example 1

Find $z_{0.05}$.

Since $\alpha = 0.050$, $z_{0.05}$ appears at the 95th percentile. That is, 95% of the scores appears below this z score and 5% appear above this z score. From a standard normal probabilities table, find the closest value to 0.9500 (0.4500 for tables showing 0.0000 to 0.5000). The value we seek lies halfway between 0.9495 and 0.9505 (0.4495 and 0.4505), corresponding to z scores of 1.64 and 1.65, respectively. Thus $z_{0.05} = 1.645$, which is halfway between 1.64 and 1.65.

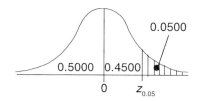

Example 2

Find $z_{0.75}$.

Since $\alpha = 0.75$, $z_{0.75}$ appears at the 25th percentile. That is, 25% of the scores appears below this z score and 75% appear above this z score. Since $z_{0.75}$ appears on the left side of the standard normal curve, it is a negative number. Because of the symmetry of the normal curve, we notice that $z_{1 - 0.75} = z_{0.25}$ has the same probability as $z_{0.75}$. Thus, we can use $z_{0.25}$, which is at the 75th percentile, to find $z_{0.75}$. In a standard normal probabilities table, the closest value to 0.7500 is 0.7486 (0.2486 for tables showing 0.000 to 0.5000), which corresponds to a z score of 0.67. Thus $z_{0.25} = 0.67$. But since $z_{0.75}$ is negative, $z_{0.75} = -0.67$.

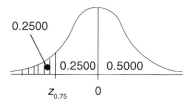

Percentile of a Normal Distribution

We can use the procedure above and standardized scores to find **probabilities of** *non*standard normal distribution.

Example 3

Applicants are admitted to a police academy only if they score in the top 60% on an entrance exam. Assume a normal distribution with $\mu = 200$ exam points and $\sigma = 50$ exam points. Find the least acceptable score to gain entrance into the academy.

We want to find $z_{0.60}$, the 40th percentile. Since $z_{0.60}$ appears on the left side of the standard normal curve, it is a negative number. Because of the symmetry of the normal curve, we notice that $z_{1-0.60} = z_{0.40}$ occurs with the same probability as $z_{0.60}$. Thus, we can use $z_{0.40}$, which is at the 60th percentile, to find $z_{0.60}$. In the standard normal distribution probabilities table, the closest value to 0.6000 is 0.5987, which corresponds to 0.25. Thus $z_{0.40} = 0.25$. But since $z_{0.60}$ is negative, $z_{0.60} = -0.25$.

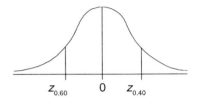

Next, we must convert $z_{0.60}$ to a nonstandard normal random variable, x. To do this, we solve

$$z = \frac{x - \mu}{\sigma} \text{ for } x:$$

◆ $x = \mu + z\sigma$

Substituting values, we have $x = 200 + (-0.25)(50) = 187.5$. So, an applicant must score at least 188 to be admitted to the academy.

?   Why do we round up the answer in the previous example?

Exercises

1. Find $z_{0.10}$.

2. Find $z_{0.95}$.

3. For a genetics experiment on carrot seeds, seeds are accepted only if the parent carrot is longer than 85% of all carrots in the geneticist's garden. This carrot population is normally distributed with a mean length of 6.7 in and a standard deviation of 0.3 in. Find the minimum parent-carrot length required for seeds to be included in the experiment.

Solutions to Exercises

Normal Probability Distributions

$0.0214 + 0.0014 = 0.0228$

Standard Normal Probability Distributions

Answers are given for standard normal probabilities tables that list values from 0.5000 to 1.0000 and from 0.0000 to 0.5000, respectively.

1. $P(z < 2.16) = 0.9846$, or $P(z < 2.16) = 0.5000 + 0.4846 = 0.9846$

2. $P(z > 2.16) = 1.0000 - 0.9846 = 0.0154$, or $P(z > 2.16) = 0.5000 - 0.4846 = 0.0154$

3. $P(-1.34 < z < 2.16) = (0.9099 - 0.5000) + (0.9846 - 0.5000) = 0.8945$,
   or $P(-1.34 < z < 2.16) = 0.4099 + 0.4846 = 0.8945$

4. $P(z < -1.85) = 1.0000 - 0.9678 = 0.0322$, or $P(z < -1.85) = 0.5000 - 0.4678 = 0.0322$

5. $P(|z| < 1.85) = P(-1.85 < z < 1.85) = (0.9678 - 0.5000) + (0.9678 - 0.5000) = 0.9356$,
   or $P(-1.85 < z < 1.85) = 0.4678 + 0.4678 = 0.9356$

Finding Normal Probabilities

1. $P(x \geq 16.3) = P\left(z \geq \dfrac{16.3 - 16.1}{0.23}\right) = P(z \geq 0.87) = 0.1922$

2. $P(x < 16.0) = P\left(z < \dfrac{16.0 - 16.1}{0.23}\right) = P(z < -0.43) = 0.3336$

3. $P(15.9 < x < 16.2) = P\left(\dfrac{15.9 - 16.1}{0.23} < z < \dfrac{16.2 - 16.1}{0.23}\right) = P(-0.87 < z < 0.43) = 0.4742$

4. $E(x + y) = 16.1 + 16.1 = 32.2$ and $SD(x + y) = \sqrt{0.23^2 + 0.23^2} = 0.33$

   So, $P(x + y < 32.5) = P\left(z < \dfrac{32.5 - 32.2}{0.33}\right) = P(z < 0.91) = 0.8186$

Percentiles of Normal Random Variables

1. $z_{0.10} = 1.28$

2. $z_{0.95} = -1.645$

3. $z_{0.15} = 1.04$. So, $1.04 = \dfrac{x - 6.7}{0.3}$, and $x = 7.01$ inches

Answers to ?

Page	Answer
6.1	A continuous random variable can theoretically take on any value within an interval of values.
6.2	0.5000
6.2	0.00 and 1.00
6.5	This new formula uses Greek letters, indicating measures for a population. Previously, we saw the formula $z = \dfrac{x - \bar{x}}{s}$, which was used to standardize data in order to compare data from different samples.
6.7	50%
6.8	Rounding down to 187 would produce a score smaller than the calculated number. Thus, 187 is an unacceptable score. We round up to 188 to ensure that the minimum score is at least as large as the calculated number.

7. Distributions of Sampling Statistics

The Central Limit Theorem

Let x be a random variable from *any* population distribution with a mean of μ and a standard deviation σ. Suppose we take a sample of size n from this population, calculate its mean, and replace the sample in the population. If we repeat this procedure several times, collecting these means, we have a **sampling distribution** of means, $\bar{x}$'s, from samples of size n. The **Central Limit Theorem** states that this sampling distribution is approximately normal with mean μ and standard deviation $\sigma/\sqrt{n}$ *regardless* of the distribution shape of the original random variable x as long as the sample size is sufficiently large. The Central Limit Theorem enables us to make an inference about a population without knowing anything about its population distribution except what we can glean from a sample.

! Generally, n must be *at least* 30 for the Central Limit Theorem to work.

? What is a parameter?

Parameters from the Central Limit Theorem

This table summarizes the parameters from the Central Limit Theorem.

	If the population distribution has:	then the sampling distribution of $\bar{x}$ has:
Random variable	x	$\bar{x}$
Mean	μ	$\mu_{\bar{x}} = \mu$
Standard deviation	σ	$\sigma_{\bar{x}} = \sigma/\sqrt{n}$

! $\mu_{\bar{x}}$ is called the **mean of the sample means**. So, $\mu_{\bar{x}} = \mu$. $\sigma_{\bar{x}}$ is called the **standard deviation of the sample means**. So, $\sigma_{\bar{x}} = \sigma/\sqrt{n}$. $\sigma_{\bar{x}}$ is also called the **standard error of the mean**.

! The Central Limit Theorem deals with probabilities about the sample mean, $\bar{x}$, not the random variable, x.

Example 1

This example shows the effect of the Central Limit Theorem on the graphs of the probability distributions of x and $\bar{x}$.

Suppose a population distribution is as shown in (a). Further suppose we graph the sampling distributions of $\bar{x}$'s, varying the sample size n. Even though the population distribution has a rectangular shape, graphs of the sampling distribution of means, (b) through (d), more closely approximate a normal curve as n increases.

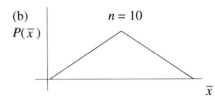

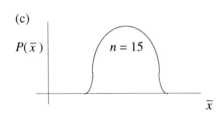

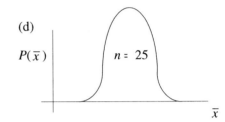

Example 2

A large, normally distributed population has a mean of 25 and a standard deviation of 10.3.

a)   Find the probability that a randomly selected score is less than 28. Note that this problem deals with the random variable, x.

$$P(x < 28) = P\left(z < \frac{28 - 25}{10.3}\right) = P(z < 0.29) = 0.6141$$

b)   If a sample of size $n = 50$ is randomly selected, find the probability that the *sample mean* is less than 28. Note that this problem deals with the sample mean, $\bar{x}$.

First, by the Central Limit Theorem, $\mu_{\bar{x}} = \mu = 25$ and $\sigma_{\bar{x}} = \dfrac{\sigma}{\sqrt{n}} = \dfrac{10.3}{\sqrt{50}} = 1.46$.

So, $P(\bar{x} < 28) = P\left(z < \dfrac{28 - 25}{1.46}\right) = P(z < 2.06) = 0.9798$

!   The Central Limit Theorem is applicable whether the random variable x is continuous or discrete, regardless of shape of the population distribution.

Sampling Without Replacement

When the sample size is large in comparison to the population size, and sampling is done *without replacement*, use

$$\sigma_{\bar{x}} = \frac{\sigma}{\sqrt{n}} \sqrt{\frac{N-n}{N-1}}$$

where σ is the population standard deviation
n is the sample size
N is the population size (that is, the population is finite)

Finite Population Correction Factor

The term $\sqrt{\dfrac{N-n}{N-1}}$ is the **finite population correction factor**.

1) When n is small in relation to N, then the finite population correction factor is near 1. So, $\sigma_{\bar{x}}$ from samples without replacement is close to $\sigma_{\bar{x}}$ from samples with replacement.

2) Generally, use this factor when $n > 0.05N$. (Some sources say $n > 0.10N$.) Otherwise, simply use $\sigma_{\bar{x}} = \dfrac{\sigma}{\sqrt{n}}$.

Summary for Standard Error of the Mean Formulas

This table summarizes which standard error formula to use depending on how you sample.

If you are sampling:	then use:
with replacement	$\sigma_{\bar{x}} = \dfrac{\sigma}{\sqrt{n}}$
without replacement and $n \leq 0.05N$ (small sample)	$\sigma_{\bar{x}} = \dfrac{\sigma}{\sqrt{n}}$
without replacement and $n > 0.05N$ (large sample)	$\sigma_{\bar{x}} = \dfrac{\sigma}{\sqrt{n}} \sqrt{\dfrac{N-n}{N-1}}$

Example 3

The mean daily food intake of 1000 laboratory animals is 20 g per animal. The standard deviation is 1.7 g per animal. Find the probability that the mean daily food intake for a random sample of 200 animals is between 19.9 and 20.25.

First, $\mu_{\bar{x}} = \mu = 20$ and since $200 \geq 0.05(1000)$,

$$\sigma_{\bar{x}} = \frac{\sigma}{\sqrt{n}}\sqrt{\frac{N-n}{N-1}} = \frac{1.7}{\sqrt{200}}\sqrt{\frac{1000-200}{1000-1}} = 0.11$$

So, $P(19.9 < \bar{x} < 20.25) = P\left(\dfrac{19.9-20}{0.11} < z < \dfrac{20.25-20}{0.11}\right) = P(-0.91 < z < 2.27) = 0.8070.$

! If the size of the population is not given in the problem, try to glean this information by the way the problem is worded. For example, if the problem concerns a national survey of some kind, you can generally assume that, unless otherwise indicated, a sampling is smaller than 0.05 of the population.

Exercises

For exercises 1-3, suppose a large population distribution has a mean of 100.1 and a standard deviation of 5.3.

1. What are the mean and standard deviation of the sample means if the sample size is 45?

2. Using a sample of size 45, what is the probability that the sample mean is between 99.5 and 100.5?

3. Two thousand golfers visit the local golf course each year. For the fifth hole, the mean number of strokes is 3.5 and the standard deviation is 0.95. Find the probability that the mean number of strokes on this hole for a random sample of 200 golfers is less than 3.6.

Normal as Approximation to the Binomial Distribution

Criteria

The normal distribution is a good approximation to the binomial distribution if and only if:
$$np \geq 5 \text{ and } nq \geq 5$$

! Make sure that *both* of the above criteria are met before using this approximation.

? What type of random variable is associated with a normal distribution? with a binomial distribution?

? What type of graph is associated with a normal distribution? with a binomial distribution?

The next example shows how to use the normal approximation to the binomial distribution.

! As before, try to identify the values of *n*, *x*, *p*, and *q* before finding the probability.

Example 1

If $n = 100$, $p = 0.20$, and $q = 0.80$, find $P(x \geq 19)$.

Since $np = 100(0.20) = 20$ *and* $nq = 100(0.80) = 80$, we can use the normal approximation to the binomial distribution.

Next, we find the mean and standard deviation for this binomial distribution:
$$\mu = np = 100(0.20) = 20 \text{ and } \sigma = \sqrt{npq} = \sqrt{100(0.20)(0.80)} = 4$$

Now standardize the random variable and determine the probability:

$P(x \geq 19) = P(x \geq 18.5)$ Correction for continuity (see below)

$= P\left(z \geq \dfrac{18.5 - 20}{4}\right)$

$= P(z \geq -0.38)$

$= 0.6480$

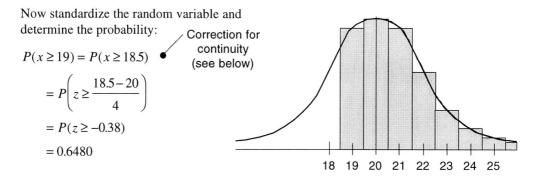

Correction for Continuity

When we use a normal distribution (continuous curve) to approximate a binomial distribution (histogram), we expand each value of the random variable x one half unit in both directions. This expansion is called the **correction for continuity**. It expands each single point to a width of one unit. This is done because, if x is a single point on a continuous curve, $P(x) = 0$. Thus, in the previous example, we use 18.5 instead of 19 as the left-hand boundary of the shaded region under the curve.

! Graphically, the correction for continuity is shown as a shaded rectangle above each point spanning a half unit to either side of the point.

Example 2

An adhesive is 50% effective in bonding two uneven metal surfaces. Find the probability that if fifteen pairs of uneven metal surfaces are selected at random, four to six pairs will be bonded by the adhesive.

First, note that $n = 15$, $x = 4$, 5, or 6, $p = 0.5$, and $q = 1 - p = 0.5$. Also, $np = 15(0.5) = 7.5 \geq 5$ *and* $nq = 15(0.5) = 7.5 \geq 5$. So, we can use the normal approximation to the binomial distribution.

We use the formulas for μ and σ for a binomial distribution:

$$\mu = np = 7.5 \text{ and } \sigma = \sqrt{npq} = 1.94$$

So,

$P(4 \leq x \leq 6)$

$= P(3.5 < x < 6.5)$

$= P\left(\dfrac{3.5 - 7.5}{1.94} < z < \dfrac{6.5 - 7.5}{1.94}\right)$

$= P(-2.07 < z < -0.52)$

$= 0.2823$

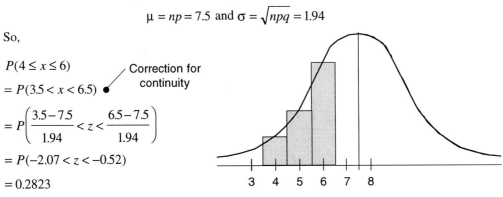

Correction for continuity

Example 3

Compare the approximating method used in example 2 to the computation using the binomial probability formula:

Note that $n = 15$, $x = 4$, 5, or 6, $p = 0.5$, and $q = 0.5$

$$P(4 \text{ bonds}) = {}_{15}C_4 \left(\frac{1}{2}\right)^4 \left(\frac{1}{2}\right)^{11} = 0.042$$

$$P(5 \text{ bonds}) = {}_{15}C_5 \left(\frac{1}{2}\right)^5 \left(\frac{1}{2}\right)^{10} = 0.092$$

$$P(6 \text{ bonds}) = {}_{15}C_6 \left(\frac{1}{2}\right)^6 \left(\frac{1}{2}\right)^9 = 0.153$$

So, $P(4 \text{ to } 6 \text{ bonds}) = 0.042 + 0.092 + 0.153 = 0.287$

Summary

We have three ways to determine a binomial probability:

- Binomial distribution probabilities table

- Binomial probability formula

- Normal approximation to the binomial distribution

Exercises

Use the normal approximation to the binomial distribution in all the following problems.

1. If $n = 35$, $p = 0.40$, and $q = 0.60$, find $P(x < 17)$.

For exercises 2 and 3, suppose that 30% of the candy bags at the SofTooth Candy Company are under the weight claimed on the package.

2. If 200 bags of candy are randomly selected, find the probability that at least 52 are underweight.

3. If 200 bags of candy are randomly selected, find the probability that between 55 and 65 are underweight.

Solutions to Exercises

The Central Limit Theorem

1. $\mu_{\bar{x}} = \mu = 100.1$ and $\sigma_{\bar{x}} = \dfrac{\sigma}{\sqrt{n}} = \dfrac{5.3}{\sqrt{45}} = 0.79$

2. $P(99.5 < \bar{x} < 100.5) = P\left(\dfrac{99.5 - 100.1}{0.79} < z < \dfrac{100.5 - 100.1}{0.79}\right) = P(-0.76 < z < 0.51) = 0.4714$

3. $\mu_{\bar{x}} = \mu = 3.5$

    Since $200 > 0.05(2000) = 100$, $\sigma_{\bar{x}} = \dfrac{\sigma}{\sqrt{n}}\sqrt{\dfrac{N-n}{N-1}} = \dfrac{0.95}{\sqrt{200}}\sqrt{\dfrac{2000-200}{2000-1}} = 0.064$

    So, $P(\bar{x} < 3.6) = P\left(z < \dfrac{3.6 - 3.5}{0.064}\right) = P(z < 1.56) = 0.9406$

Normal as Approximation to the Binomial Distribution

1. $\mu = np = 35(0.40) = 14$ and $\sigma = \sqrt{npq} = \sqrt{35(0.40)(0.60)} = 2.90$

    So, $P(x < 17) = P(x \le 16.5) = P\left(z \le \dfrac{16.5 - 14}{2.90}\right) = P(z \le 0.86) = 0.8051$

    For exercises 2 and 3, $\mu = np = 200(0.30) = 60$ and $\sigma = \sqrt{npq} = \sqrt{200(0.30)(0.70)} = 6.48$

2. $P(x \ge 52) = P(x \ge 51.5) = P\left(z \ge \dfrac{51.5 - 60}{6.48}\right) = P(z \ge -1.31) = 0.9049$

3. $P(55 \le x \le 65) = P(54.5 \le x \le 65.5) = P\left(\dfrac{54.5 - 60}{6.48} \le z \le \dfrac{65.5 - 60}{6.48}\right)$

    $= P(-0.85 \le z \le 0.85) = 0.6046$

Answers to ?

Page	Answer
7.1	A parameter is a number describing a characteristic of a population.
7.5	Continuous r. v. with a normal dist.; discrete r. v. with a binomial dist.
7.5	Bell-shaped curve with a normal dist.; histogram with a binomial dist.

8. Estimation

Estimators of a Population Mean

In this section, we look at three estimators of a population mean:
- Point estimator
- Interval estimator when the population variance is known
- Interval estimator when the population variance is unknown

Point Estimator of the Mean

When we wish to estimate the population mean by a single value, we use the sample mean, $\bar{x}$. This is called the **point estimator of the mean**.

Example 1

When a chocolate dessert is randomly selected from each of 50 restaurants in a large metropolitan area, the 50 desserts are found to have a mean of 363 calories per serving. We say that $\bar{x} = 363$ is the point estimator of μ, the average number of calories per serving in the population of chocolate desserts from all restaurants in the metropolitan area.

Background for the Interval Estimate of the Population Mean

When $\bar{x}$ is used as an estimate of μ, there will likely be an error in this estimation. We want to find a **confidence interval** based on $\bar{x}$ in which we can say μ appears with a probability of $1 - \alpha$. That is, μ will *not* appear in this interval with a probability of α. The probability $1 - \alpha$ is called the **degree of confidence**. The probability α is called the **significance level**. Our choice of a value for α depends on how confident we want to be about our interval. Generally, α is 0.01, 0.02, 0.05, or 0.10. By the figure below, the confidence interval falls between $-z_{\alpha/2}$ and $z_{\alpha/2}$. Thus, $-z_{\alpha/2}$ and $z_{\alpha/2}$ are called **critical values**. There is an $\alpha/2$ probability that μ is in either outlying region, the **left tail** or the **right tail** under the curve.

? What does the Central Limit Theorem say about the mean and standard deviation of the sampling distribution of the means?

Recall that:
- The sampling distribution of $\bar{x}$ can be approximated by a normal distribution with mean μ and standard deviation $\sigma/\sqrt{n}$.
- A z score represents the number of standard deviations a value is from the mean.

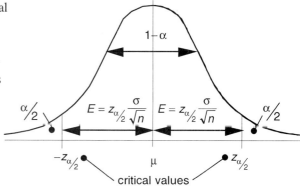

Maximum Error of the Estimate of the Mean

The **maximum error of the estimate** of μ, E, is the largest value that $\bar{x}$ can differ from μ:

$$E = z_{\alpha/2} \frac{\sigma}{\sqrt{n}}$$

Interval Estimator of the Population Mean When the Population Variance Is Known

We are $1 - \alpha$ confident that μ is in the interval

$$(\bar{x} - E,\ \bar{x} + E) \text{ which can also be written as } \bar{x} - E < \mu < \bar{x} + E$$

In either form, this is called the **confidence interval for the population mean μ.**

Example 2

Construct a 95% confidence interval for μ, the mean of the IQs of all pupils in secondary school in the country, if a sample of size 50 yields $\bar{x}$ = 118 IQ. Assume a normal distribution with σ^2 = 256.

α = 5% = 0.0500, So, 0.0500/2 = 0.0250 appears inside each of the two tails. The area outside each of the two tails is 0.5000 + 0.4750 = 0.9750. Using a standard normal probabilities table, $z_{\alpha/2} = z_{0.05/2} = z_{0.025} = 1.96$. The critical values, then, are ± 1.96.

Next, note that σ = 16, and calculate $\sigma_{\bar{x}} = \dfrac{\sigma}{\sqrt{n}} = \dfrac{16}{\sqrt{50}} = 2.26$.

Then, $E = 1.96(2.26) = 4.43$.

The confidence interval for μ is:

$(\bar{x} - E,\ \bar{x} + E)$
$= (118 - 4.43,\ 118 + 4.43)$
$= (113.57, 122.43)$

Alternately, $113.57 < \mu < 122.43$.

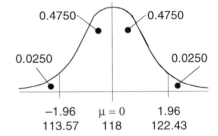

That is, we are 95% certain that μ lies between 113.57 and 122.43.

Sample Size

Sometimes we would like to know the **sample size** needed to ensure a certain level of confidence. Solving the maximum error equation for n, we have:

◆
$$n = \left(\frac{z_{\alpha/2}\sigma}{E} \right)^2$$

We can use this equation to solve problems like the following.

Example 3

We must be 98% sure that a sample of potting soil bags yields a mean weight that is within 0.5 ounces of the population mean. How large should the sample be? Assume $\sigma = 3$.

Note that $\alpha = 0.02$, $z_{\alpha/2} = z_{0.01} = 2.33$, and

$E = 0.5$. So, $n = \left(\frac{(2.33)(3)}{0.5} \right)^2 = 195.44$.

Thus, we must have a sample size of at least 196 to be 98% sure of $\bar{x}$ being within 0.5 ounces of μ.

?   Why do we always round up the answer to a problem like Example 3?

Student's *t* Distribution

If we wish to construct an interval estimator for the population mean, but the population variance in unknown, we must use the *t* distribution. The *t* distribution is very similar to the *z* distribution. If a sample from this population with unknown variance has mean $\bar{x}$ and standard deviation s, then the *t* distribution is defined by:

◆
$$t = \frac{\bar{x} - \mu}{s / \sqrt{n}}$$

?   How is the above formula different from the *z* distribution formula?

! Notice how similar the *z* distribution formula and the *t* distribution formula are to

$$z = \frac{x - \bar{x}}{s}.$$

There is a different *t* distribution for each sample size n, or, formally, for its **degrees of freedom, $n - 1$**. The number of degrees of freedom, **df**, for any statistic like *t* is equal to the number of data values that are free to vary. Use the ***t* distribution probabilities table** in your textbook to find probabilities for a *t* distribution.

Maximum Error for a t Distribution

Similar to the z distribution, we have

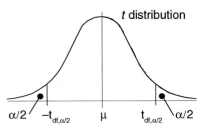

t distribution

✦
$$E = t_{df,\,\alpha/2}\,\frac{s}{\sqrt{n}}$$

!   The subscript on the critical t value shows the
degrees of freedom and the probability in each tail of the area under the curve.

Interval Estimator of the Population Mean When the Population Variance Is Unknown

The **confidence interval for a t distribution** is the same as that for a z distribution. Only
the way E is calculated has changed.

✦
$$(\bar{x}-E,\ \bar{x}+E)\ \text{which can also be written as}\ \bar{x}-E<\mu<\bar{x}+E$$

Deciding Which Distribution to Use

Use the z distribution when *either* of these conditions is met:

• σ is known and n is any size, **or**

• The sample size is $n > 30$ and σ is unknown (but estimated by s).

Use the t distribution when *all* of these conditions are met:

• The sample is small ($n \le 30$), **and**

• σ is unknown, **and**

• The population is essentially normal.

Example 4

Bags are supposed to hold 80 ounces of potting soil. A sample of 100 bags yields a mean of
$\bar{x}$ = 79 oz and a standard deviation of s = 5 oz. Construct a 98% confidence interval for μ.

Since $z_{0.02/2} = z_{0.01} = 2.33$ and $\sigma_{\bar{x}} = \dfrac{5}{\sqrt{100}} = 0.5$, then $E = 2.33(0.5) = 1.165$.

So, the 98% confidence interval for the population mean is
$(79 - 1.165,\ 79 + 1.165) = (77.835,\ 80.165)$ or $77.835 < \mu < 80.165$

?   Why do we use the z distribution in the example above?

Example 5

Twenty potting soil bags are tested for weight with these results: $\bar{x}$ = 81.5 oz and s = 5.8 oz. If σ is unknown, construct the 95% confidence interval for μ. Assume a normal distribution.

We use the t distribution. With α = 0.05 and df = 19, we find $t_{19,0.025}$ = 2.093 from a t distribution probabilities table.

We compute $E = 2.093 \left(\dfrac{5.8}{\sqrt{20}} \right) = 2.7$

So, the confidence interval is (81.5 – 2.7, 81.5 + 2.7) = (78.8, 84.2). Alternatively, 78.8 < μ < 84.2.

Thus, we are 98% confident that μ is between 78.8 and 84.2.

?   Why do we use the t distribution in the example above?

Exercises

1. What is the best point estimator of the mean length of all 1995 American-made cars if a sample yields a mean of 12.3 feet?

2. Construct a 99% confidence interval for the population mean of the volume of crates produced by the ABC Container Company. The population variance is 12.25 cubic feet. A sample of 50 from this population has a mean volume of 625.30 cubic feet.

3. Rubber bands are sold in bags marked 2 oz each. A sample of 20 bags yields a mean of 2.09 oz and a standard deviation of 0.1 oz. Construct a 98% confidence interval for the mean number of rubber bands in these bags.

4. What should the sample size be in order for us to be 98% sure that the sample mean is 1.25 units from the population mean if σ = 2.41?

Estimators of a Population Variance

In this section, we look at two estimators of a population variance:

- Point estimator
- Interval estimator

Point Estimator of the Variance

When we wish to estimate the population variance by a single value, we use the sample variance, s^2. This is called the **point estimator of the variance**.

Example 1

When a chocolate dessert is randomly selected from each of 50 restaurants in a large metropolitan area, the 50 desserts are found to have a variance of 912 calories per serving. We say that $s^2 = 912$ is the point estimator of σ^2, the variance of calories per serving in the population of chocolate desserts from all restaurants in the metropolitan area.

Background for the Interval Estimate of the Population Variance

In this section, assume the population distribution is normal. We will use the **chi-square (χ^2) distribution** with $df = n - 1$. Use the **chi-square distribution probabilities table** in your textbook to find critical χ^2 values.

! The chi-square distribution probabilities table reflects the region to the *right* of the critical values. This is unlike other probabilities tables, like the standard normal distribution table, that reflect regions to the *left* of the critical values.

We use these notations with a χ^2 distribution:

χ_L^2 = left-hand critical value

χ_R^2 = right-hand critical value

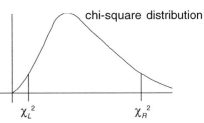
chi-square distribution

Confidence Interval for the Variance

The **confidence interval** for the population variance σ^2 is:

◆
$$\frac{(n-1)s^2}{\chi_R^2} < \sigma^2 < \frac{(n-1)s^2}{\chi_L^2}$$

For σ, we have:

◆
$$\sqrt{\frac{(n-1)s^2}{\chi_R^2}} < \sigma < \sqrt{\frac{(n-1)s^2}{\chi_L^2}}$$

! Notice that the right-hand critical value appears on the left side of the inequality and vice versa.

Example 2

Heights of all U.S. college students have $\sigma^2 = 25$ inches. It is thought that the variance of heights of students at State U. is different from the national variance. A random sample of 61 State students is measured and found to have $s^2 = 20$ inches. Determine the 95% confidence interval for σ^2.

From a χ^2 probabilities table, we find $\chi_R{}^2 = \chi_{60,0.025}^2 = 83.298$ and

$\chi_L{}^2 = \chi_{60,0.975}^2 = 40.482$.

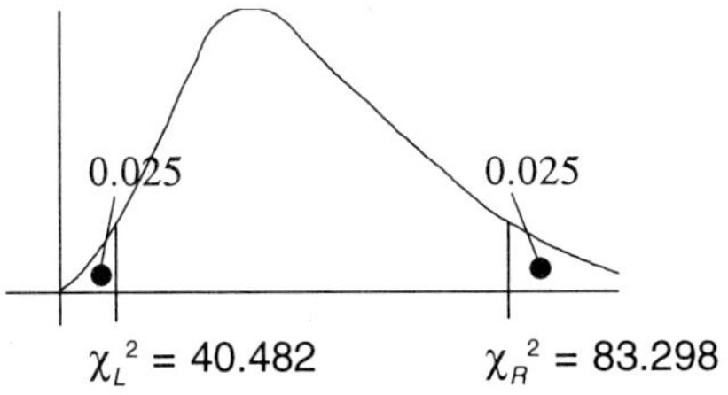

So, $\dfrac{(61-1)20}{83.298} < \sigma^2 < \dfrac{(61-1)20}{40.482}$

$14.4 < \sigma^2 < 29.6$

0.025   0.025

$\chi_L^2 = 40.482$ $\chi_R^2 = 83.298$

Thus, we are 95% certain that the population variance is between 14.4 and 29.6.

Exercises

1. What is the best point estimator of the variance for the length of all 1995 American-made cars if a sample yields a standard deviation of 1.1 feet?

2. Construct a 95% confidence interval for the population standard deviation of the volume of crates produced by the ABC Container Company. A random sample of 25 crates has a variance of 0.12 cubic feet.

Estimators of a Population Proportion

In this section, we look at two estimators of a population proportion:

- Point estimator
- Interval estimator

Point Estimator of a Population Proportion

The **population proportion** is indicated by p. The **sample proportion** is indicated by $\hat{p} = x/n$, where x is the number of successes in n trials. Also, $\hat{q} = 1 - \hat{p}$.

When we wish to estimate a population proportion, p, by a single value, use the sample proportion, $\hat{p}$. This is called the **point estimator of a population proportion**.

?   Between what values will a proportion always be?

Example 1

When a chocolate dessert is randomly selected from each of 50 restaurants in a large metropolitan area, 35 of these desserts contain over 350 calories per serving. We say that $\hat{p}$ = 35/50 = 0.70 is the point estimator of p, the population proportion of desserts containing over 350 calories per serving in the population of chocolate desserts from all restaurants in the metropolitan area.

Background for the Interval Estimate of the Population Proportion

In this section, we assume that we have a binomial distribution. Based on the formulas for the binomial distribution, we have:

✦ The **mean of the sample proportions** is $\dfrac{n\hat{p}}{n} = \hat{p}$.

✦ The **standard deviation of the sample proportions** is $\dfrac{\sqrt{n\hat{p}\hat{q}}}{n} = \sqrt{\dfrac{\hat{p}\hat{q}}{n}}$.

Maximum Error of the Estimate of the Population Proportion

The **maximum error of the estimate** of p is:

✦
$$E = z_{\alpha/2} \sqrt{\dfrac{\hat{p}\hat{q}}{n}}$$

! Notice that the z distribution, not the t distribution, is used with proportions.

Interval Estimator of a Population Proportion

The interval estimator is called the **confidence interval for the population proportion** p:

◆ $(\hat{p} - E, \hat{p} + E)$ which can also be written as $\hat{p} - E < p < \hat{p} + E$

Example 2

A car manufacturer wants to estimate the population proportion of drivers who prefer red cars. From a random sample of 200 drivers, 34 say that they prefer red cars. Find the 99% confidence interval for p.

First, $\hat{p} = 34/200 = 0.17$, $\hat{q} = 0.83$, $\alpha = 0.01$, and $z_{\alpha/2} = z_{0.01/2} = z_{0.005} = 2.575$.

So, $E = 2.575\sqrt{\dfrac{(0.17)(0.83)}{200}} = 0.07$, and the confidence interval is

$(0.17 - 0.07, 0.17 + 0.07) = (0.10, 0.24)$. Written another way, $0.10 < p < 0.24$.

So, we are 99% certain that p lies between 0.10 and 0.24.
values between 0 and 1

Sample Size

Sample size for a population proportion estimate is:

◆ $$n = z_{\alpha/2}^2 \frac{\hat{p}\hat{q}}{E^2}$$

We would use this equation to determine sample size n. But both $\hat{p} = x/n$ and $\hat{q} = 1 - \hat{p}$ depend on n and are not yet known. So, conventionally, we use $\hat{p} = 0.50$ and $\hat{q} = 0.50$, which yields a maximum value of $\hat{p}\hat{q} = 0.25$. So, we can now use:

◆ $$n = z_{\alpha/2}^2 \frac{0.25}{E^2}$$

However, an estimate, $\hat{p}$, from an earlier study may be used in the first formula for n.

? What other statistic had formulas for maximum error and sample size? What statistic did not have these formulas?

Example 3

We want to estimate, with maximum error of 0.2, the true proportion p of all college students who believe there should be no classes held on the Wednesday before Thanksgiving Day. We want a 95% confidence interval. How many students should be polled?

$\alpha = 0.05$, $z_{\alpha/2} = z_{0.025} = 1.96$, and $E = 0.2$. Since no estimate of p previously existed, we use $\hat{p} = \hat{q} = 0.50$. So,

$$n = z_{\alpha/2}^2 \frac{0.25}{E^2} = (1.96)^2 \frac{0.25}{(0.2)^2} = 24.01$$

So, we must poll 25 students to be 95% certain of coming within 0.2 of the true proportion.

Example 4

Using Example 3, suppose there had been a previous estimate of $\hat{p} = 0.70$. Find n.

$$n = (1.96)^2 \frac{(0.70)(0.30)}{(0.2)^2} = 20.17$$

So, we must poll 21 students.

Exercises

1. What is the best point estimator of the population proportion of drivers who prefer to own domestic automobiles to foreign automobiles if from a sample of 400 drivers, 254 prefer domestic vehicles?

2. Construct a 95% confidence interval for the proportion of voters who will vote for the incumbent candidate in the upcoming election. A random sample of 200 voters shows that 104 will vote for the incumbent.

3. We want to predict within 3 percentage points the proportion of voters who will vote for the incumbent candidate in the upcoming election. For a 98% confidence interval, how many voters should be polled?

4. What should the sample size be in exercise 3 if a previous poll had 134 out of 300 voters selecting the incumbent?

Solutions to Exercises

Estimators of a Population Mean

1. $\bar{x} = 12.3$ feet

2. $z_{0.005} = 2.575$, $\sigma = \sqrt{12.25} = 3.50$, $\sigma_{\bar{x}} = \dfrac{3.50}{\sqrt{50}} = 0.49$, and $E = 2.575(0.49) = 1.26$. So,

   $625.30 - 1.26 < \mu < 625.30 + 1.26$

   $624.04 < \mu < 626.56$

3. $t_{19,0.01} = 2.539$, $\sigma_{\bar{x}} = \dfrac{0.10}{\sqrt{20}} = 0.02$, and $E = 2.539(0.02) = 0.05$. So,

   $2.09 - 0.05 < \mu < 2.09 + 0.05$

   $2.04 < \mu < 2.14$

4. $E = 1.25$, $\sigma = 2.41$, $\alpha = 0.02$, and $z_{0.01} = 2.33$. So, $n = \left(\dfrac{(2.33)(2.41)}{1.25} \right)^2 = 20.18$, which rounds up to 21.

Estimators of a Population Variance

1. $s^2 = 1.21$ feet

2. $\alpha = 0.05$, $n = 25$, $df = 24$, $s^2 = 0.12$, $\chi^2_{24,0.975} = 12.401$, and $\chi^2_{24,0.025} = 39.364$. So,

   $$\sqrt{\dfrac{24(0.12)}{39.364}} < \sigma < \sqrt{\dfrac{24(0.12)}{12.401}}$$

   $0.27 < \sigma < 0.48$

Estimators of a Population Proportion

1. $\hat{p} = \dfrac{254}{400} = 0.635$

2. $\hat{p} = \dfrac{104}{200} = 0.52$, $\hat{q} = 0.48$, $a = 0.05$, $z_{0.025} = 1.96$, $E = 1.96\sqrt{\dfrac{(0.52)(0.48)}{200}} = 0.07$. So,

   $\hat{p} - E < p < \hat{p} + E$

   $0.52 - 0.07 < p < 0.52 + 0.07$

   $0.45 < p < 0.59$

3. $E = 0.03$, $a = 0.02$, and $z_{0.01} = 2.33$. So, $n = (2.33)^2 \dfrac{0.25}{(0.03)^2} = 1508.02$, which rounds up to 1509.

4. $\hat{p} = \dfrac{134}{300} = 0.45$ and $\hat{q} = 0.55$. So, $n = (2.33)^2 \dfrac{(0.45)(0.55)}{(0.03)^2} = 1492.95$, which rounds up

to 1493.

Answers to ?

Page	Answer
8.1	If the population has mean μ and standard devation σ, then sample mean is μ and sample standard deviation is $\sigma/\sqrt{n}$.
8.3	Rounding up ensures that we have at least the sample size calculated.
8.3	The z formula uses σ instead of s.
8.4	Even though σ is unknown, the sample size is sufficiently large to use the z distribution.
8.5	σ is unknown and the sample size is not sufficiently large to use the z distribution.
8.8	between 0 and 1, inclusively
8.9	Mean had these formulas, variance did not.

9. Testing Hypotheses

Basics of Hypothesis Testing

Testing hypotheses allows us to determine whether a result is statistically significant or due to chance. In this guide, we perform hypothesis tests on means, variances, and proportions. These tests are called **parametric tests** because they use parameters such as mean and variance in sampling. Each type of hypothesis test is based on a specific probability distribution, such as the normal distribution, discussed earlier in this guide.

A **hypothesis** is a statement proposing what is true in some population. The hypothesis that is assumed to be true (no effect or change) is called the **null hypothesis**, symbolized by H_0. The statement of H_0 contains relational symbols with equality: $=$, $\geq$, or $\leq$. The **alternative hypothesis**, H_1, is the negation of H_0. The statement of H_1 contains these relational symbols: $\neq$, $>$, or $<$. Relational symbols are paired in H_0 and H_1 as follows:

Symbols are correctly used if:	Example:	
H_0 contains $=$ and H_1 contains $\neq$	H_0: $\mu = 100$	H_1: $\mu \neq 100$
H_0 contains $\geq$ and H_1 contains $<$	H_0: $\mu \geq 100$	H_1: $\mu < 100$
H_0 contains $\leq$ and H_1 contains $>$	H_0: $\mu \leq 100$	H_1: $\mu > 100$

! The relational symbols we choose depend on the wording of the test scenario.

The **significance level, α,** is the probability that we have rejected H_0 when it is true.

! We generally choose α to be 0.01, 0.02, 0.05, or 0.10 depending on how certain we want to be of our hypothesis test result. For instance, we might choose $\alpha = 0.01$ when testing a hypothesis concerning a delicate surgical procedure. But we might choose $\alpha = 0.10$ when testing a claim about a sports team's progress.

Performing a Hypothesis Test

In general, all hypothesis tests come to a conclusion based on the comparison of two types of values. The first type of value is the **critical value** based on the significance level α. There are either one or two critical values depending on the hypothesis test being performed. Critical values are found either in a table based on the probability distribution or by a computer program.

? What are parameters and statistics?

The other type of value is the **test statistic** based on data, parameters, and statistics. The test statistic is calculated from a formula for the distribution for the hypothesis test. The critical values form the boundaries of the **critical regions** (also called **rejection regions**).

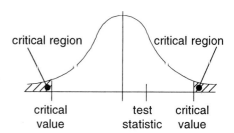

critical region   critical region

critical value   test statistic   critical value

We draw a **conclusion** as follows: If the test statistic falls in a critical region, then we reject the null hypothesis. If the test statistic does not fall in a critical region, then we fail to reject the null hypothesis.

?   What is the shape of the curve associated with the normal distribution?

Follow these steps when testing a hypothesis.

Step	Action
1	Formulate the null and alternative hypotheses based on the wording of the problem.
2	Sketch and label the distribution curve with available information.
3	Determine the critical value(s) using the significance level and include it (them) on the sketch.
4	Calculate the test statistic and mark it on the sketch.
5	Compare the test statistic to the critical value(s). With the help of the sketch, draw a conclusion about the null hypothesis.

Decision Errors

There are two **decision errors** of which to be aware:

A **type I error** occurs when we reject H_0 when it is true. $P(\text{reject } H_0 \text{ when it is true}) = \alpha$. A **type II error** occurs when we fail to reject H_0 when it is false. $P(\text{fail to reject } H_0 \text{ when it is false}) = \beta$. β cannot be determined unless the population distribution of H_1 is known.

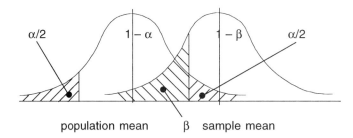

The following table summarizes correct and incorrect conclusions from a hypothesis test.

	When H_0 is true	When H_0 is false
and we reject H_0	type I error ($p = \alpha$)	correct decision ($p = 1 - \beta$)
and we fail to reject H_0	correct decision ($p = 1 - \alpha$)	type II error ($p = \beta$)

One- and Two-Tailed Tests

Hypothesis tests with two critical values are called **two-tailed**, or **non-directional**, tests. Tests with one critical value are called **one-tailed**, or **directional** tests.

Where to Place α

The relational symbol in H_1 always points in the direction of the critical region.

If H_1 contains:	the test is:	and:
$\neq$	two-tailed	α is divided evenly into each tail.
$<$	left-tailed	α is completely in the left tail.
$>$	right-tailed	α is completely in the right tail.

Testing the Mean: Known Variance

Testing the population mean when the population variance is known uses a normal distribution.

?   What are the mean and standard deviation of a sampling distribution according to the Central Limit Theorem?

Test Statistic for the Mean When Variance Is Known

Use this formula to find the **test statistic** when testing a hypothesis about the population mean and the population variance is available:

$$z = \frac{\bar{x} - \mu_{\bar{x}}}{\sigma_{\bar{x}}} = \frac{\bar{x} - \mu}{\sigma/\sqrt{n}}$$

!   As long as σ is known, sample size, n, does not affect this test. In the next section, when σ is *unknown*, n plays a part in determining which test statistic formula we use.

Example 1

This example explains more about hypothesis testing.

The mean city mileage for the DreaMobile midsize sedan is 19.5 miles per gallon with a standard deviation of 5.3 mpg. A fuel additive manufacturer wants to know whether the mean city mileage of this sedan with the fuel additive is significantly different from the city mileage without the additive. Twenty-five randomly selected DreaMobile sedans are tested with the additive resulting in a mean city mileage of 22.1 mpg. With a significance level of $\alpha = 0.05$, test the claim that the fuel additive has an effect on the sedan's mileage.

Identify what's given in the problem: $\mu = 19.5$, $\sigma = 5.3$, $\bar{x} = 22.1$, $n = 25$, and $\alpha = 0.05$.

Form the hypotheses:

H_0: $\mu = 19.5$ (The fuel additive has no effect.)

H_1: $\mu \neq 19.5$ (The fuel additive has some effect, either positive or negative.)

$\alpha = 0.05$ is split into two 0.025 areas in either tail of the normal curve since we are concerned with *both* positive and negative effects of the fuel additive (shown as the right and left tails, respectively). In this example, the two tails of the curve (0.05 of the total area under the curve) are the critical regions. If the test statistic falls into either critical region, we reject the null hypothesis. If it does not fall into either critical region, we do not reject the null hypothesis. This is a two-tailed test.

We find the z scores corresponding to the boundaries of the tails to be -1.96 and 1.96. These two z scores are the critical values.

Now, convert the observed mean, $\bar{x} = 22.1$, to a standard score. This converted score is the test statistic:

$$z = \frac{\bar{x} - \mu}{\sigma / \sqrt{n}} = \frac{22.1 - 19.5}{5.3 / \sqrt{25}} = 2.45$$

Since the test statistic falls into the right critical region (2.45 > 1.96), we reject H_0. In other words, the fuel additive had some effect on the sedan's mileage.

! The statement of the conclusion is always worded in terms of the null hypothesis. That is, we either reject H_0, or fail to reject H_0. Further, our conclusion never uses the word "accept." For example, we do not *accept* H_0 since H_0 is assumed to be true to begin with.

Example 2

Suppose that the fuel additive was thought to have not just any effect, but a *positive* effect on the sedan's mileage in Example 1. We change the hypotheses to read:

H_0: $\mu \leq 19.5$ (The additive has no effect.)
H_1: $\mu > 19.5$ (The additive improves mileage.)

This is a right-tailed test. $\alpha = 0.05$ yields a critical value of $z = 1.645$.

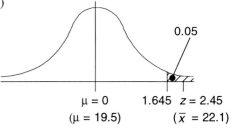

Since the test statistic $z = 2.45$ is in the critical region (that is, 2.45 > 1.645), we reject H_0. So, the fuel additive seems to have a positive effect on mileage.

Probability Value

The probability value, **p-value**, is that part of the area under the curve that is on the outside of the test statistic (*not* the critical value). The p-value is the smallest significance level at which H_0 is rejected. The smaller the p-value, the further out the test statistic is in the tail, and the greater the chance that H_0 should be rejected. A large p-value puts the test statistic closer to the population mean and increases the chance that we should fail to reject H_0.

Example 3

If the test statistic in a one-tailed test is $z = 2.03$, then using a standard normal distribution probabilities table, we find that the p-value is $1.0000 - 0.9788 = 0.0212$. So, $\alpha = 0.0212$ is the smallest significance level at which we would reject H_0.

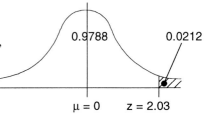

! For a two-tailed test, the p-value is the total of the areas in both tails.

Example 4

If the test statistic in a two-tailed test is $z = 2.03$, then the p-value is $0.0212 + 0.0212 = 0.0424$. So, $\alpha = 0.0424$ is the smallest significance level at which we would reject H_0.

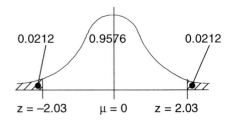

Two Ways to Draw a Conclusion About the Null Hypothesis

We can decide whether or not to reject H_0 using either of two methods:

Classical (traditional) method	p-value method
test statistic vs. critical value	p-value vs. significance level (α), or conclusion based on p-value *alone*

p-Value Versus the Significance Level

Follow these guidelines to draw a conclusion about the null hypothesis when using the p-value and α:

If	then
p-value $\leq \alpha$	reject H_0
p-value $> \alpha$	fail to reject H_0

p-Value Alone

To draw a conclusion about H_0 based on the p-value alone, follow these guidelines:

If	then
p-value < 0.01	very strongly reject H_0
$0.01 \leq p$-value ≤ 0.05	comfortably reject H_0
p-value > 0.05	strongly fail to reject H_0

Example 5

Suppose H_0: $\mu \leq 14.3$, H_1: $\mu > 14.3$, and the test statistic is $z = 1.72$. Draw a conclusion about H_0 by comparing the p-value with $\alpha = 0.05$. Draw a conclusion about H_0 by the p value alone.

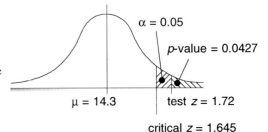

A one-tailed test with $\alpha = 0.05$ yields a critical value $z = 1.645$. The test value of $z = 1.72$ yields a p-value of 0.0427.

Comparing the p-value to α, we have $0.0427 \leq 0.05$. So, we reject H_0.

Using only the p-value, $0.01 \leq 0.0427 \leq 0.05$. So, again, we reject H_0.

Exercises

1. Each year, 7th graders at the local elementary school are given a standardized math achievement test. The results have indicated a mean score of 62.0 points and a standard deviation of 12.7 points. This year, fifty 7th graders were taught with a new math curriculum. When they took the achievement test, the 7th graders had a mean score of 59.0. At the 0.05 significance level, test the claim that the new math curriculum has affected the childrens' score on the achievement test. Use the classical method to draw a conclusion.

2. Using exercise 1, test the claim that the new math curriculum decreased the 7th graders' math scores on the achievement test. Use the classical method to draw a conclusion.

3. What is the p-value for the hypothesis test in exercise 1? in exercise 2?

4. Draw conclusions for exercises 1 and 2 based on the p-values alone.

Testing the Mean: Unknown Variance

When the population variance is unknown and sample size, n, is less than or equal to 30, we use the t distribution to perform a hypothesis test. The procedure is the same as when the population variance is known.

  When σ is unknown, but $n > 30$, substitute s for σ and use the z distribution to perform a hypothesis test. See the previous section.

Test Statistic for the Mean When Variance Is Unknown

Use this formula to find the **test statistic** when testing a hypothesis about the population mean and the population variance is unknown:

$$t = \frac{\bar{x} - \mu}{s/\sqrt{n}} \text{ with } df = n - 1$$

?   What are the criteria for using the t test?

Example 1

Consider the example from the beginning of this chapter on the fuel additive. Suppose σ is not known, but the population distribution is still normal with $\mu = 19.5$. Also, suppose that the observed mean of the experiment is $\bar{x} = 22.1$ and the standard deviation is $s = 5.1$. We are testing the claim that the fuel additive has some effect on mileage.

H_0: $\mu = 19.5$ (The additive has no effect.) H_1: $\mu \neq 19.5$ (The additive has some effect.)

Using a t distribution probabilities table, a two-tailed test with $\alpha = 0.05$ and $df = n - 1 = 25 - 1 = 24$ yields critical values of -2.064 and 2.064.
Then, the test statistic is:

$$t = \frac{\bar{x} - \mu}{s/\sqrt{n}} = \frac{22.1 - 19.5}{5.1/\sqrt{25}} = 2.55$$

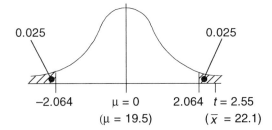

Since $2.55 > 2.064$, we reject H_0 at the 0.05 significance level. The fuel additive seems to have an effect on mileage.

Exercises

1. A beverage company claims that each of its 12-oz cans of soft drink contains one gram of sugar. A sample of ten soft drink cans yields a mean of 1.2 grams of sugar and a standard deviation of 0.4 g. Is the company's claim about the sugar content in their product correct. Use $\alpha = 0.01$.

2. Using exercise 1, test the hypothesis that the company is putting more sugar in their soft drink than they claim.

Testing a Population Proportion

In this section, we test hypotheses concerning a population's proportions, percentages, and rates. We use a normal distribution as an approximation to a binomial distribution.

! The words probability, percentage, and proportion have basically the same meaning in this section.

Test Statistic for a Proportion Test

Use either of these formulas for the **test statistic** when testing the population proportion.

$$z = \frac{x - \mu}{\sigma} = \frac{x - np}{\sqrt{npq}} \quad \text{or} \quad z = \frac{\hat{p} - p}{\sqrt{\dfrac{pq}{n}}}$$

? What do each of the symbols in the above formulas represent?

! Write down the information given in the problem to help you with the solution.

Example 1

A radio talk show hosted by a long-time media personality attracts 41% of listeners at the time this show is on the air. This host resigns and a talk show host from another city is the replacement. A survey is conducted by the station management to determine whether the new host has changed the proportion of listeners. It is found that 72 out of 200 radio listeners during this time tune in to the talk show. Test the hypothesis that the proportion of listeners remains unchanged at $\alpha = 0.05$.

$H_0: p = 0.41$ and $H_1: p \neq 0.41$

This is a binomial distribution. (People either listen or do not listen to this show.) We identify the information given to us:

$n = 200$	$x = 72$
$p = 0.41$	$q = 0.59$

$\hat{p} = 72/200 = 0.36$

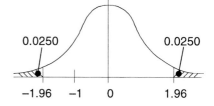

0.0250   0.0250

−1.96   −1   0   1.96

✔ All proportions are between 0 and 1.

Next, we check to see if we can use the normal approximation to the binomial:

$np = 200(0.41) = 82 \geq 5$ and $nq = 100(0.59) = 59 \geq 5$

So, we can use the normal approximation to the binomial with $\mu = np = 82$, $\sigma^2 = npq = 48.38$, and $\sigma = 6.96$.

Next, we find the critical values and the test statistic, as with other types of hypothesis tests.

The critical values of $z = \pm 1.96$

The test statistic can be found in either of two ways:

$$z = \frac{x - \mu}{\sigma} = \frac{72 - 82}{6.96} = -1.44 \quad \text{or} \quad z = \frac{\hat{p} - p}{\sqrt{\dfrac{pq}{n}}} = \frac{0.36 - 0.41}{\sqrt{\dfrac{(0.41)(0.59)}{200}}} = -1.44$$

So, since $-1.96 < -1.44 < 1.96$, we fail to reject H_0. That is, the new host neither increases nor decreases the proportion of listeners.

Exercises

1. A beverage company claims that each of its 12-oz cans of soft drink contains 10% real fruit juice. A sample of 500 soft drink cans yields 7.6% real fruit juice. At the 0.05 significance level, test the claim that the company is putting 10% real fruit juice in their soft drink.

2. Using exercise 1, test the hypothesis that the company is putting less real fruit juice in their soft drink than they claim.

Testing the Standard Deviation or Variance

In this section, we test the standard deviation and variance using the chi-square distribution. The population from which we are sampling *must be* normally distributed.

? Under what other circumstances did we use the chi-square distribution?

Test Statistic for a Standard Deviation or Variance Test

Use this formula for the **test statistic** when testing the population standard deviation or variance:

◆
$$\chi^2 = \frac{(n-1)s^2}{\sigma^2}$$

The chi-square **critical value** for this hypothesis test has $df = n - 1$.

Example 1

Test the hypothesis that electricians have a significantly higher dispersion in annual wages than the general population in a certain city. A sample of 30 electricians is polled. It is found that the variance of their annual wages is $\sigma^2 = \$14,000$. The general population has a variance of $12,000. Use $\alpha = 0.01$.

$H_0: \sigma^2 \leq 12,000$

$H_1: \sigma^2 > 12,000$

From a chi-square distribution probabilities table with $\alpha = 0.01$ and $df = 29$, the critical χ^2 value is 49.588.

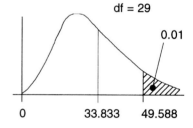

The test value is $\chi^2 = \frac{(n-1)s^2}{\sigma^2} = \frac{(29)(14,000)}{12,000} = 33.833$.

Since $33.833 < 49.588$, we do not reject H_0. That is, electricians in this city have the same variance in annual wages as the general population.

Example 2

Using example 1, test the hypothesis that electricians have a significantly different dispersion in annual wages than the general population in a certain city.

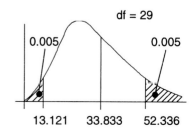

$H_0: \sigma^2 = 12,000$

$H_1: \sigma^2 \neq 12,000$

With $\alpha = 0.01$ and $df = 29$, the right-hand critical value is $\chi_R^2 = \chi_{29,0.005}^2 = 52.336$. The left-hand critical value is $\chi_L^2 = \chi_{29,0.995}^2 = 13.121$.

Since $13.121 < 33.833 < 52.336$, we do not reject H_0. The electricians in this city have the same variance in annual wages as the general population.

Exercises

1. A gardener wants a variety of sizes in a single type of daisy. In past years, the variance in the diameter of the daisies has been 1.1 cm. This year, the gardener applies a new plant food to 25 plants and notes that the variance in the diameter of the daisies is 1.3 cm. Does the new plant food affect the variance in the diameter of the daisies at $\alpha = 0.05$?

2. Using exercise 1, does the new plant food increase the variability of the daisies' diameter?

Solutions to Exercises

Testing the Mean: Known Variance

1. $H_0: \mu = 62$ $H_1: \mu \neq 62$

   Critical values are $z = \pm 1.96$. Test statistic is $z = \dfrac{59-62}{12.7 / \sqrt{50}} = -1.67$.

   Conclusion: since $-1.96 < -1.67 < 1.96$, we do not reject H_0. That is, the math curriculum does not affect the children's scores.

2. $H_0: \mu \geq 62$ $H_1: \mu < 62$

   Critical value is $z = -1.645$. Test statistic is $z = \dfrac{59-62}{12.7 / \sqrt{50}} = -1.67$.

   Conclusion: since $-1.67 < -1.645$, we reject H_0. That is, the math curriculum seems to have affected the children's scores.

3. For exercise 1, a two-tailed test and the test statistic $z = -1.67$ yield p-value $= 0.0475 + 0.0475 = 0.0950$. For exercise 2, a one-tailed test and the test statistic $z = -1.67$ yield p-value $= 0.0475$.

4. For exercise 1, p-value $= 0.0950$ is relatively large (> 0.05) and indicates that we should not reject H_0. For exercise 2, p-value $= 0.0475$ is in a range (between 0.01 and 0.05) where we can comfortably reject H_0.

Testing the Mean: Unknown Variance

1. $H_0: \mu = 1.0$ $H_1: \mu \neq 1.0$

   Critical values are $t = \pm 3.250$. Test statistic is $t = \dfrac{1.2-1.0}{0.4 / \sqrt{10}} = 1.58$.

   Conclusion: since $-3.250 < 1.58 < 3.250$, we do not reject H_0. That is, the company is not misrepresenting the sugar content.

2. $H_0: \mu \leq 1.0$ $H_1: \mu > 1.0$

   Critical value is $t = 2.821$. Test statistic is $t = \dfrac{1.2-1.0}{0.4 / \sqrt{10}} = 1.58$.

   Conclusion: since $1.58 < 2.821$, we do not reject H_0. That is, the company is not putting more sugar in their soft drink than they claim.

Testing a Population Proportion

1. $H_0: p = 0.10$ $H_1: p \neq 0.10$

    Critical values are $z = \pm 1.96$. Test statistic is $z = \dfrac{0.076 - 0.100}{\sqrt{\dfrac{(0.10)(0.90)}{500}}} = -1.79$.

    Conclusion: since $-1.96 < -1.79 < 1.96$, we do not reject H_0. That is, the company is not misrepresenting its claim.

2. $H_0: p \geq 0.10$ $H_1: p < 0.10$

    Critical value is $z = -1.645$. Test statistic is $z = \dfrac{0.076 - 0.100}{\sqrt{\dfrac{(0.10)(0.90)}{500}}} = -1.79$.

    Conclusion: since $-1.79 < -1.645$, we reject H_0. That is, there seems to be less fruit juice than claimed.

Testing the Standard Deviation or Variance

1. $H_0: \sigma^2 = 1.1$ $H_1: \sigma^2 \neq 1.1$

    Critical values are $\chi^2_{24,0.975} = 12.401$ and $\chi^2_{24,0.025} = 39.364$.

    Test statistic is $\chi^2 = \dfrac{(24)(1.3)}{1.1} = 28.36$.

    Conclusion: since $12.401 < 28.36 < 39.364$, we do not reject H_0. That is, the plant food does not affect the variance of daisies' diameter.

2. $H_0: \sigma^2 \leq 1.1$ $H_1: \sigma^2 > 1.1$

    Critical value is $\chi^2_R = \chi^2_{24,0.05} = 36.415$.

    Test statistic is $\chi^2 = \dfrac{(24)(1.3)}{1.1} = 28.36$.

    Conclusion: since $28.36 < 36.415$, we do not reject H_0. That is, the plant food does not increase the variance of daisies' diameter.

Answers to ?

Page	Answer
9.1	Parameters are numbers that describe a characteristic of a population. Statistics are numbers that describe a characteristic of a sample.
9.2	bell-shaped curve
9.4	If the population has mean m and standard deviation s, then the sample has mean m and standard deviation $\sigma/\sqrt{n}$.
9.8	see Deciding Which Distribution to Use, page 8.4
9.9	see Point Estimator of a Population Proportion, page 8.8

10. Hypothesis Tests for Two Populations

Hypothesis Tests for Two Population Parameters

In this chapter, we test whether the difference between *two* sample means, variances, or proportions indicates that the samples were taken from two different populations or from the same population. If the samples are from the same population, the difference is due to chance. The procedure we use for hypothesis testing in this chapter is the same as that in the previous chapter.

> **?** What two measurements did we include in the hypotheses from the previous chapter?

As in the previous chapter, test statistic formulas for means and proportions follow the form of the standardization formula, $z = \dfrac{x - \mu}{\sigma}$. In this chapter, the null hypothesis states that the two population parameters are equal. That is, the difference between the parameters is 0. For example, $H_0: \mu_1 = \mu_2$ or $H_0: \mu_1 - \mu_2 = 0$. Consequently, test statistic formulas for comparing two population parameters do not include the "μ part" of the standardization formula.

> **?** How are these hypotheses different from the previous chapter?

Independent and Dependent Samples

Two samples are **independent** if they are made up of two distinct groups of subjects. For **dependent samples**, each observation from one sample is paired or related in some way with an observation in another sample.

Types of Hypothesis Tests for Two Population Parameters

In this chapter, there is only one hypothesis test for two population variances and only one for two population proportions. However, there are *five* kinds of hypothesis tests for two population means:

Samples	Population variances	Sample sizes	Assumptions about population variances	Distribution
dependent				t
independent	known			z
independent	unknown	large		z
independent	unknown	small	assumed equal	F and t
independent	unknown	small	assumed unequal	F and t

We discuss the particulars of these tests next.

Paired Sample *t* Test for Two Means

Testing Dependent Samples

Let $\bar{d}$ be the mean value of the difference, d, between x and y, where x and y are paired observations from samples taken from two normal dependent populations with means μ_1 and μ_2 and standard deviations σ_1 and σ_2, respectively. For n pairs of observations, the values of d will have a t distribution with mean $\mu_d = \mu_1 - \mu_2$ and standard deviation $s_d / \sqrt{n}$, where s_d is the standard deviation of all d values for paired observations. The **test statistic** is:

$$t = \frac{\bar{d} - \mu_d}{s_d / \sqrt{n}}$$

The **critical values** have $df = n - 1$

The **maximum error** of the estimate of the population mean is:

$$E = t_{df, \alpha/2} \frac{s_d}{\sqrt{n}} \quad \text{with } df = n - 1$$

The **confidence interval** for the mean is:

$$\bar{d} - E < \mu_d < \bar{d} + E$$

Example - Hypothesis Test

We wish to determine whether students' achievements in music are different from their achievements in mathematics. A sample of 8 students is randomly selected with grades as follows:

Music x	Mathematics y	Difference $d = x - y$
79	80	−1
83	75	8
88	85	3
93	96	−3
98	88	10
58	65	−7
75	65	10
68	73	−5
		15

 Pairs of grades are dependent since they are from the same student.

Test the hypothesis that average achievement in music is the same as average achievement in mathematics at $\alpha = 0.05$.

We state the hypotheses in either of two equivalent ways:

$H_0: \mu_1 = \mu_2$ $H_1: \mu_1 \neq \mu_2$

or $H_0: \mu_1 - \mu_2 = 0$ $H_1: \mu_1 - \mu_2 \neq 0$

!   We assume that $\mu_d = 0$ since H_0: $\mu_d = \mu_1 - \mu_2 = 0$

At $\alpha = 0.05$ and $df = 8 - 1 = 7$, the critical values are
$t = \pm 2.365$.

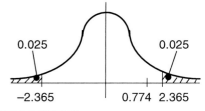

Calculate $\bar{d} = \dfrac{15}{8} = 1.875$.

Using the formula variance for ungrouped data, $s_d = \sqrt{\dfrac{n \Sigma d^2 - (\Sigma d)^2}{n(n-1)}} = 6.854$.

The test statistic is $t = \dfrac{\bar{d} - \mu_d}{s_d / \sqrt{n}} = \dfrac{1.875 - 0}{6.854 / \sqrt{8}} = 0.774$.

Since $-2.365 < 0.774 < 2.365$, we fail to reject H_0. The average achievement in music is the same as average achievement in mathematics.

Example - Confidence Interval

Construct the 95% confidence interval for these data.

First, calculate $E = t_{7, 0.025} \dfrac{s_d}{\sqrt{n}} = 2.365 \cdot \dfrac{6.854}{\sqrt{8}} = 5.727$

Then, the 95% confidence interval is:

$$\bar{d} - E < \mu_d < \bar{d} + E$$
$$1.875 - 5.727 < \mu_d < 1.875 + 5.727$$
$$-3.852 < \mu_d < 7.602$$

That is, we are 95% certain that μ_d is between -3.852 and 7.602.

Exercises

1. In order to reduce the number of defective parts, management has established quality assurance training for workers making the parts. The number of defective parts before and after training are shown in the table. At $\alpha = 0.01$, test the hypothesis that training has reduced the number of defective parts.

Worker	Before training	After training
1	5	3
2	5	6
3	4	4
4	7	5

2. Construct the 99% confidence interval for the data in exercise 1.

Testing Two Means: Known Variances

Testing Independent Samples

For *independent* samples, tests about the difference between two means (like tests about a single mean) are made under two categories:

1. Population variances, σ_1^2 and σ_2^2, are known.
2. Population variances are not known and must be estimated by s_1^2 and s_2^2 taken from two large samples (covered in the next section).

?   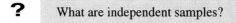 What are independent samples?

When Population Variances Are Known

When both population variances are known, use this **test statistic**:

$$z = \frac{(\bar{x}_1 - \bar{x}_2) - (\mu_1 - \mu_2)}{\sqrt{\dfrac{\sigma_1^2}{n_1} + \dfrac{\sigma_2^2}{n_2}}}$$

!   We state $H_0: \mu_1 = \mu_2$. That is, $\mu_1 - \mu_2 = 0$ and, as mentioned before, $\mu_1 - \mu_2$ can be omitted from the above formula.

The **maximum error** of the estimate of the population mean is:

$$E = z_{\alpha/2} \sqrt{\frac{\sigma_1^2}{n_1} + \frac{\sigma_2^2}{n_2}}$$

The **confidence interval** for the mean is:

$$(\bar{x}_1 - \bar{x}_2) - E < \mu_1 - \mu_2 < (\bar{x}_1 - \bar{x}_2) + E$$

!   When constructing a confidence interval, do not assume $\mu_1 - \mu_2 = 0$. It is only in a hypothesis test that we make this assumption.

Example 1 - Hypothesis Test

We want to determine whether the average caffeine content of a cola from one soft drink company is equal to that of a cola from a second soft drink company. The variances of the two colas are $\sigma_1^2 = 38$ mg and $\sigma_2^2 = 42$ mg. Samples of 100 colas from each bottler are randomly selected. The means are found to be $x_1 = 18$ mg and $\bar{x}_2 = 20$ mg. Test whether the colas have the same mean caffeine content at $\alpha = 0.05$.

$H_0: \mu_1 = \mu_2$ (or $\mu_1 - \mu_2 = 0$)

$H_1: \mu_1 \neq \mu_2$ (or $\mu_1 - \mu_2 \neq 0$)

For $\alpha/2 = 0.025$, we have the critical values
of $z = \pm1.96$.

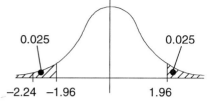

The test statistic is $z = \dfrac{\bar{x}_1 - \bar{x}_2}{\sqrt{\dfrac{\sigma_1^{2}}{n_1} + \dfrac{\sigma_2^{2}}{n_2}}} = \dfrac{18 - 20}{\sqrt{\dfrac{38}{100} + \dfrac{42}{100}}} = -2.24$

Since $-2.24 < -1.96$, we reject H_0. The caffeine content in the first cola is not the same as that in the second cola.

Further, the **p-value** is 0.025. So, H_0 would be rejected at any significance level greater than or equal to 0.025.

Example 1 - Confidence Interval

Construct the 95% confidence interval for these data.

First, calculate $E = z_{\alpha/2} \sqrt{\dfrac{\sigma_1^{2}}{n_1} + \dfrac{\sigma_2^{2}}{n_2}} = 1.96\sqrt{\dfrac{38}{100} + \dfrac{42}{100}} = 1.75$. Then, the 95% confidence interval is:

$$(\bar{x}_1 - \bar{x}_2) - E < \mu_1 - \mu_2 < (\bar{x}_1 - \bar{x}_2) + E$$

$$(18 - 20) - 1.75 < \mu_1 - \mu_2 < (18 - 20) + 1.75$$
$$-3.75 < \mu_1 - \mu_2 < -0.25$$

We are 95% certain that the difference between the two population means is between -3.75 and -0.25.

Example 2

Two makes of trucks are compared for braking ability. Random samples of 75 trucks from each of the two populations are selected. Distances required to stop at 60 mph are observed. Suppose that $\sigma_1^{2} = 200$ feet for the first make of truck and $\sigma_2^{2} = 224$ feet for the second make of truck. The means for the samples are $\bar{x}_1 = 155$ feet and $\bar{x}_2 = 160$ feet. Does the first make of truck stop in a significantly shorter distance than the second make of truck at $\alpha = 0.01$?

$H_0: \mu_1 \geq \mu_2$ (or $\mu_1 - \mu_2 \geq 0$)

$H_1: \mu_1 < \mu_2$ (or $\mu_1 - \mu_2 < 0$)

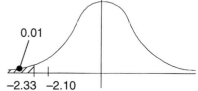

At $\alpha = 0.01$, the critical value is $z = -2.33$.

The test statistic is $z = \dfrac{155 - 160}{\sqrt{\dfrac{200}{75} + \dfrac{224}{75}}} = -2.10$.

Since $-2.33 < -2.10$, we fail to reject H_0. The first make of truck stops in a significantly shorter distance than the second make of truck.

Exercises

1. Two neighborhoods have the same city services. However, there is concern that neighborhood A homeowners pay more in real estate taxes than neighborhood B homeowners. Samples of real estate taxes from 50 households in each neighborhood yield means of $3500 and $3470 for the A and B neighborhoods, respectively. The variances in real estate taxes for the neighborhoods are $\sigma^2 = \$4700$ for A and $\sigma^2 = \$4600$ for B. Do the two neighborhoods pay the same amount of taxes for the same city services at $\alpha = 0.05$?

2. What is the p-value of test statistic in exercise 1?

3. Construct the 95% confidence interval for the data in exercise 1.

Testing Two Means: Unknown Variances

When Population Variances Are Unknown

To test a hypothesis about two population means when population variances are unknown, we use s_1^2 and s_2^2 to estimate σ_1^2 and σ_2^2.

! *Both* sample sizes, n_1 and n_2, must be greater than 30 in this case.

The **test statistic** in this case is:

✦
$$z = \frac{\bar{x}_1 - \bar{x}_2}{\sqrt{\dfrac{s_1^2}{n_1} + \dfrac{s_2^2}{n_2}}}$$

The **maximum error** of the estimate of the population mean is:

✦
$$E = z_{\alpha/2}\sqrt{\frac{s_1^2}{n_1} + \frac{s_2^2}{n_2}}$$

The corresponding **confidence interval** is:

✦
$$(\bar{x}_1 - \bar{x}_2) - E < \mu_1 - \mu_2 < (\bar{x}_1 - \bar{x}_2) + E$$

? What is the difference between these formulas and those used when population variances are known?

Example - Hypothesis Test

A group of dairy farmers wish to increase the milk yield from their cows through diet. Part of the group feeds their herds diet I, the other part feeds their herds diet II. A sample of 100 cows from each population is randomly selected. The first sample has a mean daily increase in milk yield of 10 lbs and a standard deviation of 13 lbs. The second sample has a mean daily increase in milk yield of 7 lbs and a standard deviation of 9 lbs. Test H_0 that the two diets have the same effect on milk yield at $\alpha = 0.01$.

$H_0: \mu_1 = \mu_2$

$H_1: \mu_1 \neq \mu_2$

0.005   0.005

−2.575   1.90 2.575

The population variances are unknown and $n_1 = n_2 > 30$. At $\alpha = 0.01$ and two-tailed test, we have critical values of $z = \pm 2.575$. The test

statistic is $z = \dfrac{10 - 7}{\sqrt{\dfrac{13^2}{100} + \dfrac{9^2}{100}}} = 1.90$

Since $-2.575 < 1.90 < 2.575$, we do not reject H_0. The two diets have the same effect on milk yield.

Further, the **p-value** is 0.0574. So, H_0 would be rejected at any significance level greater than or equal to 0.0574.

Example - Confidence Interval

Construct the 99% confidence interval for these data.

First, calculate $E = z_{\alpha/2} \sqrt{\dfrac{s_1^2}{n_1} + \dfrac{s_2^2}{n_2}} = 2.575 \sqrt{\dfrac{169}{100} + \dfrac{81}{100}} = 4.07$. Then, the 99% confidence interval is:

$$(\bar{x}_1 - \bar{x}_2) - E < \mu_1 - \mu_2 < (\bar{x}_1 - \bar{x}_2) + E$$

$$(10 - 7) - 4.07 < \mu_1 - \mu_2 < (10 - 7) + 4.07$$

$$-1.07 < \mu_1 - \mu_2 < 7.07$$

We are 99% certain that the difference between the two population means is between -1.07 and 7.07.

Exercises

1. Two neighborhoods have the same city services. However, there is concern that neighborhood A homeowners pay more in real estate taxes than neighborhood B homeowners. Samples of real estate taxes from 50 households in each neighborhood yield means of $3500 and $3470 and standard deviations of $4550 and $4470 for the A and B neighborhoods, respectively. Does neighborhood A pay more real estate taxes than neighborhood B for the same city services at $\alpha = 0.02$?

2. Construct the 98% confidence interval for the data in exercise 1.

Testing Two Population Variances

In this section, we compare population variances by using a new distribution. We assume that the populations in this section are normally distributed.

F Distribution

The **F distribution** is used for testing whether or not the variance of one population is equal to that of another. It uses *two* degrees of freedom. The **test statistic** is:

◆
$$F = \frac{s_1{}^2}{s_2{}^2}$$

! In the F test statistic, $s_1{}^2 \geq s_2{}^2$. That is, the numerator is *never* less than the denominator. If one sample variance is larger than the other, you must use that sample variance as the numerator. Thus, $F \geq 1$ always, and we need be concerned only with the *right-hand* critical value when testing variances.

If the sample variances are equal, $F = 1$ and we do not reject H_0. Generally, $s_1{}^2$ and $s_2{}^2$ are somewhat different. The F distribution enables us to determine whether this difference is due to chance or is large enough for us to reject H_0. Use an **F distribution probabilities table** to find the **critical value**, F_{df_1, df_2, α_R}, where $df_1 = n_1 - 1$, $df_2 = n_2 - 1$, and α_R is the area under the curve in the *right* tail.

! $s_1{}^2$, n_1, and df_1 come from the same population, and $s_2{}^2$, n_2, and df_2 come from the same population.

Example 1

We wish to determine whether test scores in mathematics taught with cooperative learning techniques are more variable than test scores in mathematics taught by the lecture method. A sample is drawn from a population of high school students taught with cooperative learning and another sample is drawn from a population of students taught by lecturing. A sample of 25 test scores taken from the first population yields a variance of $s_1{}^2 = 35$. Another sample of 20 observations taken from the second population yields a variance of $s_2{}^2 = 26$. If $\alpha = 0.05$, do test scores have a greater variability when mathematics is taught with cooperative learning techniques than by the lecture method?

$H_0\colon \sigma_1{}^2 \leq \sigma_2{}^2$

$H_1\colon \sigma_1{}^2 > \sigma_2{}^2$

A one-tailed test, $\alpha = 0.05$, $df_1 = 25 - 1 = 24$, $df_2 = 20 - 1 = 19$ yields a critical value of

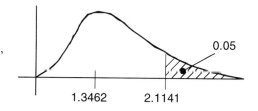

0.05

1.3462   2.1141

$F_{24,19,0.05} = 2.1141$. The test statistic is $F = \dfrac{s_1^{\,2}}{s_2^{\,2}} = \dfrac{35}{26} = 1.3462$.

✔ Test statistic ≥ 1

Since $1.3462 < 2.1141$, we fail to reject H_0. The test scores do not have a greater variability when mathematics is taught with cooperative learning techniques than by the lecture method.

Example 2

Two robots are used to produce rods 200 centimeters in length, though the lengths may vary slightly in the production process. The company engineers believe that the variability of the rod lengths produced by one robot differs from the variability of the rod lengths produced by the other robot. Random samples of 29 and 31 rods are selected from the respective robots. The samples yield variances of 0.5 cm and 0.6 cm, respectively. Test the hypothesis that the population variances of the rods produced by the two robots are equal at $\alpha = 0.05$.

$H_0: \sigma_1^2 = \sigma_2^2$

$H_1: \sigma_1^2 \neq \sigma_2^2$

Since $s_1^2 = 0.6$ is greater than $s_2^2 = 0.5$, $df_1 = 31 - 1 = 30$ and $df_2 = 29 - 1 = 28$.

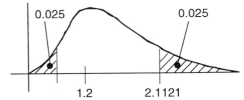

The critical value is $F_{30,28,0.025} = 2.1121$. The test statistic is $F = \dfrac{s_1^{\,2}}{s_2^{\,2}} = \dfrac{0.6}{0.5} = 1.2$.

Since $1.2 < 2.1121$, we do not reject H_0. The robots produce rods with the same population variance.

Exercise

City planners believe that the variability in real estate taxes paid in neighborhood A is greater that the variability of real estate taxes paid in neighborhood B. A sample of real estate taxes from 41 households in neighborhood A yields a variance of $850. A sample of real estate taxes from 61 households in neighborhood B yields a variance of $970. Are the real estate taxes paid by neighborhood A more variable than the taxes paid by neighborhood B? Use $\alpha = 0.05$.

Testing Two Means: Small Samples

Suppose the population variances are unknown. If either n_1 or $n_2 \leq 30$, we cannot use the normal distribution. As before, we must use the t distribution instead. There are two cases involved here:

1. The two populations have *equal* variances.
2. The two populations have *unequal* variances.

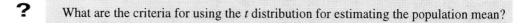

? What are the criteria for using the t distribution for estimating the population mean?

We determine whether variances are equal or not by performing a hypothesis test with the F distribution.

Equal Variances with Small Samples

If, through the F distribution, we can assume that the population variances are equal, the **test statistic** is:

$$t = \frac{\overline{x}_1 - \overline{x}_2}{\sqrt{\frac{(n_1 - 1)s_1{}^2 + (n_2 - 1)s_2{}^2}{n_1 + n_2 - 2}\left(\frac{1}{n_1} + \frac{1}{n_2}\right)}}$$

The **critical values** have $df = n_1 + n_2 - 2$

The **maximum error** of the estimate of the population mean is:

$$E = t_{df,\,\alpha/2}\sqrt{\frac{(n_1 - 1)s_1{}^2 + (n_2 - 1)s_2{}^2}{n_1 + n_2 - 2}\left(\frac{1}{n_1} + \frac{1}{n_2}\right)} \quad \text{with } df = n_1 + n_2 - 2$$

! $\dfrac{(n_1 - 1)s_1{}^2 + (n_2 - 1)s_2{}^2}{n_1 + n_2 - 2}$ is called the **pooled estimator of σ^2**.

The **confidence interval** for the population mean is:

$$(\overline{x}_1 - \overline{x}_2) - E < \mu_1 - \mu_2 < (\overline{x}_1 - \overline{x}_2) + E$$

Example 1 - Hypothesis Test

We wish to determine whether teaching mathematics with cooperative learning techniques and teaching mathematics by the lecture method have the same effect on learning. Two classes of math students are selected at random. One class is taught with cooperative learning, the other class is taught by lecture. At the end of the semester, the two classes take a standardized exam. The cooperative learning group of 12 students had an average score of $\overline{x} = 84$ and a variance of $s^2 = 14$. The lecture group of 13 students had an average score of $\overline{x} = 79$ and a variance of $s^2 = 18$. Test H_0 that both methods are equally effective at $\alpha = 0.05$.

We conduct a preliminary test using the F distribution to determine whether or not we can assume the population variances are equal.

$H_0: \sigma_1^2 = \sigma_2^2$ $H_1: \sigma_1^2 \neq \sigma_2^2$

The critical value is $F_{12,11,0.025} = 3.4296$.

The test statistic is $F = s_1^2/s_2^2 = 18/14 = 1.2857$. Since $1.2857 < 3.4296$, we fail to reject H_0. That is, we can assume that $\sigma_1^2 = \sigma_2^2$ and use the t statistic above.

Now, to the main problem.

Hypotheses are $H_0: \mu_1 - \mu_2 = 0$ and $H_1: \mu_1 - \mu_2 \neq 0$

At $\alpha = 0.05$ and $df = n_1 + n_2 - 2 = 12 + 13 - 2 = 23$, the critical values are t = ±2.069. The test

statistic is $t = \dfrac{79 - 84}{\sqrt{\dfrac{(12)(18) + (11)(14)}{13 + 12 - 2}\left(\dfrac{1}{13} + \dfrac{1}{12}\right)}} = -3.115$.

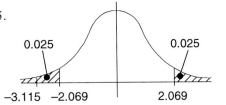

0.025   0.025

−3.115 −2.069   2.069

Since $-3.115 < -2.069$, we reject H_0. Both methods are equally effective.

Example 1 - Confidence Interval

Construct the 95% confidence interval for these data.

First, calculate $E = 2.069\sqrt{\dfrac{(12)(18) + (11)(14)}{13 + 12 - 2}\left(\dfrac{1}{13} + \dfrac{1}{12}\right)} = 3.321$. Then, the 95% confidence

interval is:

$$(\bar{x}_1 - \bar{x}_2) - E < \mu_1 - \mu_2 < (\bar{x}_1 - \bar{x}_2) + E$$
$$(79 - 84) - 3.321 < \mu_1 - \mu_2 < (79 - 84) + 3.321$$
$$-8.321 < \mu_1 - \mu_2 < -1.679$$

Unequal Variances with Small Samples

If, through the F distribution, we can assume that the population variances are *not* equal, the **test statistic** is:

$$t = \dfrac{\bar{x}_1 - \bar{x}_2}{\sqrt{\dfrac{s_1^2}{n_1} + \dfrac{s_2^2}{n_2}}}$$

The **critical values** have df = lesser of $n_1 - 1$ and $n_2 - 1$.

The **maximum error** of the estimate of the population mean is:

$$E = t_{df,\alpha/2}\sqrt{\dfrac{s_1^2}{n_1} + \dfrac{s_2^2}{n_2}} \quad \text{with } df = \text{lesser of } n_1 - 1 \text{ and } n_2 - 1.$$

Example 2

Consider Example 1 but suppose the cooperative learning group of 12 students had an average score of $\bar{x} = 84$ and a variance of $s^2 = 5$. The lecture group of 13 students had an average score of $\bar{x} = 79$ and a variance of $s^2 = 18$.

The preliminary test to determine whether or not we can assume the two population variances are equal at $\alpha = 0.05$ is as follows.

$H_0: \sigma_1^2 = \sigma_2^2$ $\qquad\qquad$ $H_1: \sigma_1^2 \neq \sigma_2^2$

The critical value is $F_{12,11,0.025} = 3.4296$.

The test statistic is $F = s_1^2/s_2^2 = 18/5 = 3.6000$. Since $3.6000 > 3.4296$, we reject H_0. That is, we can assume that $\sigma_1^2 \neq \sigma_2^2$ and use the t statistic above.

Now, to the main problem.

$H_0: \mu_1 - \mu_2 = 0$

$H_1: \mu_1 - \mu_2 \neq 0$

0.025 $\qquad\qquad\qquad\qquad$ 0.025

-3.726 −2.201 $\qquad\qquad$ 2.201

At $\alpha = 0.05$ and $df = 12 - 1 = 11$ (smaller than $13 - 1 = 12$), the critical values are $t = \pm 2.201$.

The test statistic is $t = \dfrac{79 - 84}{\sqrt{\dfrac{18}{13} + \dfrac{5}{12}}} = -3.726$.

Since $-3.726 < -2.201$, we reject H_0. Both methods are not equally effective.

Exercises

1. Two neighborhoods have the same city services. However, there is concern that neighborhood A homeowners pay more in real estate taxes than neighborhood B homeowners. Random samples of real estate taxes from the two neighborhoods are taken with results as shown. Does neighborhood A pay more real estate taxes than neighborhood B for the same city services at $\alpha = 0.05$?

Neighborhood	Sample size	Mean	Variance
A	10	$3510	$4720
B	12	$3450	$4690

2. Perform the hypothesis test in exercise 1, if the sample variances are $6400 for neighborhood A and $2100 for neighborhood B.

3. Construct the 95% confidence interval for the data in exercise 1.

Testing Two Population Proportions

In this section, we will test whether two population proportions, p_1 and p_2, are equal or differ significantly.

The sampling distribution of the difference between two sample proportions, $\hat{p}_1 - \hat{p}_2$, is approximately normally distributed, if n_1 and n_2 are sufficiently large. It has mean $p_1 - p_2$ and

standard deviation $\sigma_{\hat{p}_1 - \hat{p}_2} = \sqrt{\dfrac{p_1 q_1}{n_1} + \dfrac{p_2 q_2}{n_2}}$.   (1)

Pooled Estimate of p_1 and p_2

In practical situations, p_1, q_1, p_2, and q_2 are generally unknown. But, if H_0 assumes $p_1 = p_2$, then we can pool the data from both samples to obtain estimates of p_1, q_1, p_2, and q_2. This is called the **pooled estimate of p_1 and p_2** and is written

◆
$$\overline{p} = \frac{x_1 + x_2}{n_1 + n_2}$$ (2)

Also,

◆
$$\overline{q} = 1 - \overline{p}$$ (3)

Substituting the estimates $\overline{p}$ and $\overline{q}$ into (1) for p_1, p_2, q_1, and q_2, we have

$\sigma_{\hat{p}_1 - \hat{p}_2} = \sqrt{\overline{p}\,\overline{q}\left(\dfrac{1}{n_1} + \dfrac{1}{n_2}\right)}$.

Test Statistic and Maximum Error of the Estimate

The **test statistic** to be used for comparing two proportions is:

◆
$$z = \frac{(\hat{p}_1 - \hat{p}_2) - (p_1 - p_2)}{\sqrt{\overline{p}\,\overline{q}\left(\dfrac{1}{n_1} + \dfrac{1}{n_2}\right)}}$$

❗ We state $H_0: p_1 = p_2$. That is, $H_0: p_1 - p_2 = 0$. So, $p_1 - p_2$ and can be omitted from the above formula.

The **maximum error** of the estimate of the population proportion is:

◆
$$E = z_{\alpha/2} \sqrt{\frac{p_1 q_1}{n_1} + \frac{p_2 q_2}{n_2}}$$

The **confidence interval** for the population proportion is:

◆
$$(\hat{p}_1 - \hat{p}_2) - E < p_1 - p_2 < (\hat{p}_1 - \hat{p}_2) + E$$

Example 1 - Hypothesis Test

A scientist wants to determine whether the proportion of persons in two populations with connected earlobes is the same. Random samples from each of the two populations are taken and the number in each population with this characteristic is noted. A sample of 50 from the first population has 12 persons with the characteristic. A sample of 60 from the second population has 16 persons with the characteristic. Test H_0 that the population proportions are the same at $\alpha = 0.05$.

$H_0: p_1 = p_2$ (or $p_1 - p_2 = 0$) and $H_1: p_1 \neq p_2$ (or $p_1 - p_2 \neq 0$)

The critical values are $z = \pm 1.96$.

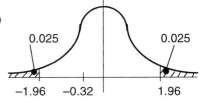

From the problem, we know $\hat{p}_1 = \dfrac{x_1}{n_1} = \dfrac{12}{50} = 0.240$,

$\hat{p}_2 = \dfrac{x_2}{n_2} = \dfrac{16}{60} = 0.267$, $\bar{p} = \dfrac{12+16}{50+60} = 0.255$, and $\bar{q} = 1 - 0.255 = 0.745$.

So, the test value is $z = \dfrac{0.240 - 0.267}{\sqrt{(0.255)(0.745)\left(\dfrac{1}{50} + \dfrac{1}{60}\right)}} = -0.32$

Since $-1.96 < -0.32 < 1.96$, we fail to reject H_0. The population proportions are the same.

Further, the **p-value** is 0.749, which is large enough so that H_0 would never be rejected at any reasonable significance level.

!   The letter p is used in two different ways here: p_1 and p_2 are population proportions, and p-value is the probability value used to draw a conclusion about the null hypothesis.

Example 1 - Confidence Interval

Construct the 95% confidence interval for these data.

First, calculate $E = z_{\alpha/2} \sqrt{\dfrac{\hat{p}_1 \hat{q}_1}{n_1} + \dfrac{\hat{p}_2 \hat{q}_2}{n_2}} = 1.96 \sqrt{\dfrac{(0.240)(0.760)}{50} + \dfrac{(0.267)(0.733)}{60}} = 0.163$.

Then the 95% confidence interval is:

$$(\hat{p}_1 - \hat{p}_2) - E < p_1 - p_2 < (\hat{p}_1 - \hat{p}_2) + E$$
$$(0.240 - 0.267) - 0.163 < p_1 - p_2 < (0.240 - 0.267) + 0.163$$
$$0.190 < p_1 - p_2 < 0.136$$

We are 95% certain that the difference between population proportions is between -0.190 and 0.136.

Example 2

It is believed that the proportion of voters in the 1st ward who will vote in the upcoming primary exceeds the proportion of voters in the 2nd ward who will vote in the same primary. A survey is taken with these results:

Ward	n	Number planning to vote
1st	$n_1 = 330$	$x_1 = 256$
2nd	$n_2 = 350$	$x_2 = 238$

Test the above hypothesis with $\alpha = 0.05$.

$H_0: p_1 \le p_2$ (or $p_1 - p_2 \le 0$)

$H_1: p_1 > p_2$ (or $p_1 - p_2 > 0$)

The critical value is $z = 1.645$.

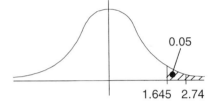

From the problem, we know $p_1 = \dfrac{x_1}{n_1} = \dfrac{256}{330} = 0.776$, $p_2 = \dfrac{x_2}{n_2} = \dfrac{238}{350} = 0.680$,

$\bar{p} = \dfrac{256+238}{330+350} = 0.726$, and $\bar{q} = 1 - 0.726 = 0.274$.

So, the test value is $z = \dfrac{0.776 - 0.680}{\sqrt{(0.726)(0.274)\left(\dfrac{1}{330} + \dfrac{1}{350}\right)}} = 2.74$.

Since $2.74 > 1.645$, we reject H_0. The proportion of voters in the first ward exceeds the proportion of voters in the second ward.

Exercises

1. Homeowners in each of two neighborhoods are polled to determine how many believe they are paying more real estate taxes than the other neighborhood. Random samples of homeowners from the two neighborhoods are taken with results as shown. Test the claim that the same proportion of homeowners in the two neighborhoods think they pay more taxes than the other neighborhood at $\alpha = 0.05$?

Neighborhood	Sample size	Number thinking they pay more
A	25	15
B	30	16

2. Construct the 95% confidence interval for the data in exercise 1.

Solutions to Exercises

Paired Sample *t* Test for Two Means

1. $H_0: \mu_1 \leq \mu_2$ and $H_1: \mu_1 > \mu_2$, $\bar{d} = \dfrac{3}{4} = 0.75$, and $s_d = 1.5$. Critical value is $t_{3,0.01} = 4.541$.

   Test statistic is $t = \dfrac{0.75}{1.5 \big/ \sqrt{4}} = 1.0$. Since $1.0 < 4.541$, we fail to reject H_0. Training has not

   reduced the number of defective parts.

2. Critical value is $t_{3,0.005} = 5.841$ and $E = 5.841 \cdot \dfrac{1.5}{\sqrt{4}} = 4.3808$. So, the 99% confidence

   interval is: $0.75 - 4.3808 < \mu_d < 0.75 + 4.3808$
   $$-3.6308 < \mu_d < 5.1308$$

Testing Two Means: Known Variances

1. $H_0: \mu_1 = \mu_2$ and $H_1: \mu_1 \neq \mu_2$. The critical values are $z = \pm 1.96$. The test statistic is

   $z = \dfrac{3500 - 3470}{\sqrt{\dfrac{4700}{50} + \dfrac{4600}{50}}} = \dfrac{30}{13.64} = 2.20$. Since $2.20 > 1.96$, we reject H_0. The two

   neighborhoods do not pay the same amount of taxes for the same city services.

2. The *p*-value is 0.0278.

3. $E = 1.96\sqrt{\dfrac{4700}{50} + \dfrac{4600}{50}} = 26.73$. So, the 95% confidence interval is

   $(3500 - 3470) - 26.73 < \mu_1 - \mu_2 < (3500 - 3470) + 26.73$
   $$3.27 < \mu_1 - \mu_2 < 56.73$$

Testing Two Means: Unknown Variances

1. $H_0: \mu_1 \leq \mu_2$ and $H_1: \mu_1 > \mu_2$. Since $\alpha = 0.02$, the critical value is $z = 2.05$. The test statistic

   is $z = \dfrac{3500 - 3470}{\sqrt{\dfrac{4550}{50} + \dfrac{4470}{50}}} = 2.23$. Since $2.23 > 2.05$, we reject H_0. Neighborhood A pays

   more taxes than neighborhood B for the same city services.

2. Using $\alpha/2 = 0.01$, $E = 2.33\sqrt{\dfrac{4550}{50} + \dfrac{4470}{50}} = 31.29$. So, the 98% confidence interval is

   $(3500 - 3470) - 31.29 < \mu_1 - \mu_2 < (3500 - 3470) + 31.29$
   $$-1.29 < \mu_1 - \mu_2 < 61.29$$

Testing Two Population Variances

$H_0: \sigma_1^2 \leq \sigma_2^2$ and $H_1: \sigma_1^2 > \sigma_2^2$. The critical value is $F_{60,40,0.05} = 1.6373$. The test statistic is

$$F = \frac{s_1^2}{s_2^2} = \frac{970}{850} = 1.1412.$$ Since $1.1412 < 1.6373$, we fail to reject H_0. Taxes paid by

neighborhood A are not more variable than the taxes paid by neighborhood B.

Testing Two Means: Small Samples

1. Preliminary test of $H_0: \sigma_1^2 = \sigma_2^2$ and $H_1: \sigma_1^2 \neq \sigma_2^2$:
    The critical value is $F_{9,11,0.025} = 3.5879$. The test statistic is $F = s_1^2/s_2^2 = 4720/4690 = 1.0064$. Since $1.0064 < 3.5879$, we fail to reject H_0, that is, assume $\sigma_1^2 = \sigma_2^2$.

    Main test of $H_0: \mu_1 \leq \mu_2$ and $H_1: \mu_1 > \mu_2$.
    Critical value is $t_{20,0.05} = 1.725$. Test statistic is

    $$t = \frac{3510 - 3450}{\sqrt{\dfrac{(9)(4720) + (11)(4690)}{10 + 12 - 2}\left(\dfrac{1}{10} + \dfrac{1}{12}\right)}} = 2.043.$$ Since $2.043 > 1.725$, we reject H_0.

    Neighborhood A pays more taxes than neighborhood B.

2. Preliminary test of $H_0: \sigma_1^2 = \sigma_2^2$ and $H_1: \sigma_1^2 \neq \sigma_2^2$:
    The critical value is $F_{9,11,0.05} = 2.8962$. The test statistic is $F = s_1^2/s_2^2 = 6400/2100 = 3.0476$. Since $3.0476 > 2.8962$, we reject H_0, that is, assume $\sigma_1^2 \neq \sigma_2^2$.

    Main test of $H_0: \mu_1 \leq \mu_2$ and $H_1: \mu_1 > \mu_2$:

    Critical value is $t_{9,0.05} = 1.833$. Test statistic is $t = \dfrac{3510 - 3450}{\sqrt{\dfrac{6400}{10} + \dfrac{2100}{12}}} = 2.102.$

    Since $2.102 > 1.833$, we reject H_0. Neighborhood A pays more taxes than neighborhood B.

3. $E = 2.086 \sqrt{\dfrac{(9)(4720) + (11)(4690)}{10 + 12 - 2}\left(\dfrac{1}{10} + \dfrac{1}{12}\right)} = 61.26.$ Then, the 95% confidence interval is

    $$(3510 - 3450) - 61.26 < \mu_1 - \mu_2 < (3510 - 3450) + 61.26$$
    $$-1.26 < \mu_1 - \mu_2 < 121.26$$

Testing Two Population Proportions

1. $H_0: p_1 = p_2$ and $H_1: p_1 \neq p_2$

    $\hat{p}_1 = \dfrac{15}{25} = 0.6$, $\hat{p}_2 = \dfrac{16}{30} = 0.53$, $\bar{p} = \dfrac{15+16}{25+30} = 0.56$, and $\bar{q} = 0.44$. The critical values are

    $z = \pm 1.96$. The test statistic is $z = \dfrac{0.60 - 0.53}{\sqrt{(0.56)(0.44)\left(\dfrac{1}{25} + \dfrac{1}{30}\right)}} = 0.52$.

    Since $-1.96 < 0.52 < 1.96$, we do not reject H_0. The same proportion of homeowners in the two neighborhoods think they pay more taxes that the other neighborhood.

2. $E = 1.96\sqrt{\dfrac{(0.60)(0.40)}{25} + \dfrac{(0.53)(0.47)}{30}} = 0.26$. Then the 95% confidence interval is

    $$(0.60 - 0.53) - 0.26 < p_1 - p_2 < (0.60 - 0.53) + 0.26$$
    $$-0.19 < p_1 - p_2 < 0.33$$

Answers to ?

Page	Answer
10.1	population parameter (e.g., mean) and a statistic from a sample of that population.
10.1	Hypotheses in this chapter compare two parameters from two populations.
10.4	Two samples are independent if they are made up of two distinct groups of subjects, unrelated in any way.
10.7	s is used as an estimate of σ.
10.11	see Deciding Which Distribution to Use, page 8.4

11. Analysis of Variance

One-Factor ANOVA with Equal Sample Sizes

In this section, we test the null hypothesis that the differences among *several sample means* are due to chance. That is, these sample means are all equal. We assume normal distributions, approximately equal variances, and independent samples. We use *two* estimates of the population variance in an F distribution to do this. This is a method called **one-factor analysis of variance (one-factor ANOVA)**.

Test Statistic for Equal Sample Sizes

For tests involving *equal sample sizes*, the **test statistic** is:

$$\blacklozenge \qquad F = \frac{\text{variance between samples}}{\text{variance within samples}} = \frac{ns_{\bar{x}}^2}{\left(\Sigma s^2\right)/m}$$

where n = number of values in each sample
m = number of samples

$s_{\bar{x}}^2$ = variance of the sample means (first estimate of σ^2)

$\left(\Sigma s^2\right)/m$ = mean of the sample variances (second estimate of σ^2)

The **critical value** has degrees of freedom $df_1 = m - 1$ from the numerator of F, and degrees of freedom $df_2 = m(n - 1)$ from the denominator of F.

Example

Distinct techniques are used in teaching German to three groups of randomly selected students. At the end of the semester, the students are given a standardized test with these results.

	Technique		
	1	2	3
	x_1	x_2	x_3
	77	76	84
	74	78	86
	75	85	82
	73	82	89
	$\bar{x}_1 = 74.8$	$\bar{x}_2 = 80.3$	$\bar{x}_3 = 85.3$
	$s_1^2 = 2.9$	$s_2^2 = 16.3$	$s_3^2 = 8.9$

$n = 4$ scores in each sample
$m = 3$ samples

?   Explain how the means and variances in the table above were calculated.

Test the claim that the three techniques are equally effective at $\alpha = 0.05$. That is, we want to determine whether differences among the sample means are sufficiently large or due to chance.

Our hypotheses are:
$H_0: \mu_1 = \mu_2 = \mu_3$
H_1: At least one mean is different than the other means.

Degrees of freedom are $df_1 = m - 1 = 2$ and $df_2 = m(n - 1) = 3(3) = 9$ with $\alpha = 0.05$. So, the critical value is $F_{2,9,0.05} = 4.2565$.

To determine the test statistic, we find the two estimates of the population variance. First,

$$s_{\bar{x}}^2 = \frac{m(\Sigma \bar{x}^2) - (\Sigma \bar{x})^2}{m(m-1)} = \frac{3(74.8^2 + 80.3^2 + 85.3^2) - (74.8 + 80.3 + 85.3)}{3(2)} = 27.6$$

Then, $\dfrac{\Sigma s^2}{m} = \dfrac{s_1^2 + s_2^2 + s_3^2}{3} = \dfrac{2.9 + 16.3 + 8.9}{3} = 9.4$

So, the test value is $F = \dfrac{4(27.6)}{9.4} = 11.7$.

Since $11.7 > 4.2565$, we reject H_0.

Exercise

Three brands of microwave ovens are sampled to determine their life spans. The results are shown in the table. Do the three brands have equal life spans? Use $\alpha = 0.05$.

Microwave life spans		
1	2	3
x_1	x_2	x_3
12	10	13
8	9	11
9	11	9
7	12	12
8	10	10
$\bar{x}_1 = 8.8$	$\bar{x}_2 = 10.4$	$\bar{x}_3 = 11.0$
$s_1^2 = 3.7$	$s_2^2 = 1.3$	$s_3^2 = 2.5$

One-Factor ANOVA with Unequal Sample Sizes

This section covers one-factor ANOVA for *unequal* sample sizes. Again, we use *two* estimates of the population variance in an *F* distribution.

?   Other than ANOVA, where did we use the *F* distribution?

Test Statistic for Unequal Sample Sizes

For tests involving *unequal sample sizes*, the **test statistic** is:

$$F = \frac{\text{variance between samples}}{\text{variance within samples}} = \frac{\left[\dfrac{\Sigma n_i (\bar{x}_i - \bar{\bar{x}})^2}{m-1}\right]}{\left[\dfrac{\Sigma (n_i - 1)s_i^2}{N-m}\right]}$$

where $\bar{\bar{x}}$ = overall mean (mean of all sample means)
m = number of samples
N = total number of data in all samples

The **critical value** has degrees of freedom $df_1 = m - 1$ from the numerator of *F*, and degrees of freedom $df_2 = N - m$ from the denominator of *F*.

The numerator of the test statistic is called the **mean square for treatment**. It, in turn, is the **sum of squares for treatment** divided by $m - 1$.

$$MS(\text{treatment}) = \frac{SS(\text{treatment})}{m-1}$$

The denominator of the test statistic is called the **mean square for error**. It, in turn, is the **sum of squares for error** divided by $N - m$.

$$MS(\text{error}) = \frac{SS(\text{error})}{N-m}$$

Example

Suppose the data from the example in the previous section were as follows with unequal sample sizes. Test the claim that the three techniques are equally effective. Use $\alpha = 0.05$.

	Technique		
1	2	3	
x_1	x_2	x_3	
77	76	84	
74	78	86	$m = 3$ samples
75	85	82	
73		89	
82			
$\bar{x}_1 = 76.2$	$\bar{x}_2 = 79.7$	$\bar{x}_3 = 85.3$	
$s_1^2 = 12.7$	$s_2^2 = 22.3$	$s_3^2 = 8.9$	
$\Sigma x = 381$	$\Sigma x = 239$	$\Sigma x = 341$	$\Sigma\Sigma x = 961$
$n = 5$	$n = 3$	$n = 4$	$N = 12$

The degrees of freedom for the numerator are $m - 1 = 3 - 1 = 2$. The degrees of freedom for the denominator are $N - m = 12 - 3 = 9$. So, the critical value is $F_{2,9,0.05} = 4.2565$.

The overall mean is $\bar{\bar{x}} = \dfrac{961}{12} = 80.1$.

The variance between samples (numerator of the test statistic) is

$$\frac{\Sigma n_i (\bar{x}_i - \bar{\bar{x}})^2}{m-1} = \frac{5(76.2-80.1)^2 + 3(79.7-80.1)^2 + 4(85.3-80.1)^2}{3-1} = 92.4 .$$

The variance within samples (denominator of the test statistic) is

$$\frac{\Sigma (n_i - 1)s_i^2}{N-m} = \frac{(5-1)(12.7)+(3-1)(22.3)+(4-1)(8.9)}{12-3} = 13.6$$

So, the test statistic is $F = \dfrac{92.4}{13.6} = 6.8$. Since $6.8 > 4.2565$, we reject H_0.

Exercise

Three brands of microwave ovens are sampled to determine their life span. The results are shown in the table. Do the three brands have equal life spans? Use $\alpha = 0.05$.

Microwave life spans		
1	2	3
x_1	x_2	x_3
12	10	13
8	9	11
9	11	10
7	12	
8	10	
	12	
$\bar{x}_1 = 8.8$	$\bar{x}_2 = 10.7$	$\bar{x}_3 = 11.3$
$s_1^2 = 3.7$	$s_2^2 = 1.5$	$s_3^2 = 2.3$

Two-Factor ANOVA

In the previous sections of this chapter, we considered the effect of a single factor on samples of our data. In this section, we look at the effect of *two non-interacting* factors on a single sample of our data. We assume a normal distribution, a common variance σ^2, and an independent sample. This method is called **two-factor analysis of variance (two-factor ANOVA)**.

We are interested in two hypothesis tests. One test claims that there is no effect on the data from the first of the two factors (**row factor**):

H_0: No effect from the row factor

The other test claims that there is no effect on the data from the second of the two factors (**column factor**):

H_0: No effect from the column factor

As with one-factor ANOVA, we use *two* estimates of the population variance in an F distribution to perform each of these tests.

! This is a very calculation-intensive method and is typically done with a computer.

Tabulating the Data

Sample data is arranged in m rows and n columns. The rows show data regarding one factor. The columns show data regarding the other factor. The data value in the ith row and jth column is shown as x_{ij}. For example, x_{21} represents the data value in row 2, column 1.

Next, we compute a mean for each row and each column. We compute the **overall mean** for all data and place it in the lower right corner of the table. Notice the use of **dot notation** to express these means. For example, $x_{1\bullet}$ is the mean of the data in the first row, $x_{\bullet 2}$ is the mean of the data in the second column, and $x_{\bullet\bullet}$ is the overall mean.

	Column Factor				Mean
	x_{11}	x_{12}	...	x_{1n}	$x_{1\bullet}$
	x_{21}	x_{22}	...	x_{2n}	$x_{2\bullet}$
Row factor	...	...	...	...	...
	x_{m1}	x_{m2}	...	x_{mn}	$x_{m\bullet}$
Mean	$x_{\bullet 1}$	$x_{\bullet 2}$	...	$x_{\bullet n}$	$x_{\bullet\bullet}$

Double Summation

The next topic involves a formula with a double summation. A double summation has this form:

$$\sum_{i=1}^{m} \sum_{j=1}^{n} x_{ij}$$

i and j are called indices (plural of **index**). Index i has integral values from 1 through m. Index j has integral values from 1 through n. To evaluate this double summation, start with the rightmost index j. Vary the value of index j from 1 through n for each value of index i (1 through m). Add the resulting data.

Example 1

Evaluate $\sum\limits_{i=1}^{2}\sum\limits_{j=1}^{3} x_{ij}$.

$$\sum_{i=1}^{2}\sum_{j=1}^{3} x_{ij} = x_{11} + x_{12} + x_{13} + x_{21} + x_{22} + x_{23} + x_{31} + x_{32} + x_{33}$$

First Estimate of σ

This first estimate of σ is always valid regardless of whether or not the null hypothesis is true.

◆
$$\frac{SS_e}{N} = \frac{\sum\limits_{i=1}^{m}\sum\limits_{j=1}^{n}\left(x_{ij} - x_{i\bullet} - x_{\bullet j} + x_{\bullet\bullet}\right)^2}{(n-1)(m-1)}$$

! SS_e is called the **error sum of squares**.

Second Estimate of σ

This second estimate of σ is valid only when the null hypothesis is true. There are two versions of this estimator depending on whether you are testing for row effect or column effect. In both cases m = number of rows and n = number of columns.

When testing the row factor, use:

◆
$$\frac{SS_r}{m-1} = \frac{n\sum\limits_{i=1}^{m}\left(x_{i\bullet} - x_{\bullet\bullet}\right)^2}{m-1}$$

! SS_r is called the **row sum of squares**.

When testing the column factor, use:

◆
$$\frac{SS_c}{n-1} = \frac{m\sum\limits_{j=1}^{n}\left(x_{\bullet j} - x_{\bullet\bullet}\right)^2}{n-1}$$

! SS_c is called the **column sum of squares**

Test Statistic

The **test statistic** consists of the estimators discussed above.

When testing the row factor, use

$$\blacklozenge \qquad F = \frac{SS_r/(m-1)}{SS_e/N}.$$

The **critical value** has $df_1 = m - 1$ and $df_2 = N = (m - 1)(n - 1)$.

When testing the column factor, use

$$\blacklozenge \qquad F = \frac{SS_c/(n-1)}{SS_e/N}.$$

The **critical value** has $df_1 = n - 1$ and $df_2 = N = (m - 1)(n - 1)$.

Example 2

The following are the number of a certain bacteria per cm² grown under two different amounts of light and three temperatures. We assume there is no interaction between the light and temperature factors. We arrange the data in a 2 x 3 table, shown in the center of the figure below. The rows show the data with regard to the first factor, the amount of light. The columns show the data with regard to the second factor, the temperature. We wish to test these two null hypotheses at the 0.05 significance level:

H_0: Light has no effect on bacteria growth. (No row effect.)
H_0: Temperature has no effect on bacteria growth. (No column effect.)

Next, we compute a mean for each row and each column and the overall mean.

| | | Temperatures (in degrees Celsius) | | | |
		20	25	30	Mean
Light (in	500	1400	1550	1700	$x_{1\bullet} = 1550$
lumens)	750	1550	1650	1750	$x_{2\bullet} = 1650$
Mean		$x_{\bullet 1} = 1475$	$x_{\bullet 2} = 1600$	$x_{\bullet 3} = 1725$	$x_{\bullet\bullet} = 1600$

The row sum of squares is:

$$SS_r = n \sum_{i=1}^{m} (x_{i\bullet} - x_{\bullet\bullet})^2 = 3\left[(1550-1600)^2 + (1650-1600)^2\right] = 15000$$

The column sum of squares is:

$$SS_c = m \sum_{j=1}^{n} (x_{\bullet j} - x_{\bullet\bullet})^2 = 2\left[(1475-1600)^2 + (1600-1600)^2 + (1725-1600)^2\right] = 62500$$

The two previous results serve in the numerators of the two test statistics. The error sum of squares is part of the denominator of these test statistics:

$$SS_e = \sum_{i=1}^{2} \sum_{j=1}^{3} (x_{ij} - x_{i\bullet} - x_{\bullet j} + x_{\bullet\bullet})^2$$

$$= (x_{11} - x_{1\bullet} - x_{\bullet 1} + x_{\bullet\bullet})^2 + (x_{12} - x_{1\bullet} - x_{\bullet 2} + x_{\bullet\bullet})^2 + (x_{13} - x_{1\bullet} - x_{\bullet 3} + x_{\bullet\bullet})^2$$

$$+ (x_{21} - x_{2\bullet} - x_{\bullet 1} + x_{\bullet\bullet})^2 + (x_{22} - x_{2\bullet} - x_{\bullet 2} + x_{\bullet\bullet})^2 + (x_{23} - x_{2\bullet} - x_{\bullet 3} + x_{\bullet\bullet})^2$$

$$= (1400 - 1550 - 1475 + 1600)^2 + (1550 - 1550 - 1600 + 1600)^2 + (1700 - 1550 - 1725 + 1600)^2$$

$$+ (1550 - 1650 - 1475 + 1600)^2 + (1650 - 1650 - 1600 + 1600)^2 + (1750 - 1650 - 1725 + 1600)^2$$

$$= 625 + 0 + 625 + 625 + 0 + 625 = 2500$$

Before we test the two hypotheses, note that $N = (m - 1)(n - 1) = (2 - 1)(3 - 1) = 2$.

The test statistic for the hypothesis testing the row factor is:

$$F = \frac{\text{second estimate of } \sigma}{\text{first estimate of } \sigma} = \frac{SS_r / (m-1)}{SS_e / N} = \frac{15000 / (2-1)}{2500 / 2} = 12.000 .$$

The critical value is $F_{1,2,0.05} = 18.513$. Since $12.000 < 18.513$, we conclude that light has no significant effect on the number of bacteria.

The test statistic for the hypothesis testing the column factor is:

$$F = \frac{\text{second estimate of } \sigma}{\text{first estimate of } \sigma} = \frac{SS_c / (n-1)}{SS_e / N} = \frac{62500 / (3-1)}{2500 / 2} = 25.000 .$$

The critical value is $F_{2,2,0.05} = 19.000$. Since $25.000 > 19.000$, we conclude that temperature has a significant effect on the number of bacteria.

Exercise

Average lifespans (in years) of a certain type of female mammal are shown below. Test whether daily vitamin C supplements throughout the mammals' lives and the average number of litters over their lifetimes affect the lifespan of these female mammals. Use a = 0.025.

		Vitamin C supplements (in milligrams)	
		50	100
Number	1	18	19
of	5	17	19
litters	10	10	12

Solutions to Exercises

One-Factor ANOVA with Equal Sample Sizes

H_0: $\mu_1 = \mu_2 = \mu_3$ H_1: At least one mean is different than the other means.

$df_1 = m - 1 = 2$ and $df_2 = m(n-1) = 3(4) = 12$. The critical value is $F_{2,12,0.05} = 3.8853$.

The first estimate of the population variance (numerator) is

$$s_{\bar{x}}^2 = \frac{3(8.8^2 + 10.4^2 + 11.0^2) - (8.8 + 10.4 + 11.0)^2}{3(2)} = 1.3$$

The second estimate (denominator) is $\dfrac{\Sigma s^2}{m} = \dfrac{3.7 + 1.3 + 2.5}{3} = 2.5$

So, the test value is $F = \dfrac{5(1.3)}{2.5} = 2.6$

Since $2.6 < 3.8853$, we do not reject H_0.

One-Factor ANOVA with Unequal Sample Sizes

$df_1 = m - 1 = 3 - 1 = 2$. $df_2 = N - m = 14 - 3 = 11$. So, the critical value is $F_{2,11,0.05} = 3.9823$.

The overall mean is $\bar{\bar{x}} = \dfrac{142}{14} = 10.1$.

The variance between samples (numerator of the test statistic) is:

$$\frac{\Sigma n_i (\bar{x}_i - \bar{\bar{x}})^2}{m-1} = \frac{5(8.8 - 10.1)^2 + 6(10.7 - 10.1)^2 + 3(11.3 - 10.1)^2}{3 - 1} = 7.5$$

The variance within samples (denominator of the test statistic) is:

$$\frac{\Sigma (n_i - 1)s_i^2}{N - m} = \frac{(5-1)(3.7) + (6-1)(1.5) + (3-1)(2.3)}{14 - 3} = 2.4$$

So, the test statistic is $F = \dfrac{7.5}{2.4} = 3.1$. Since $3.1 < 3.9823$, we do not reject H_0.

Two-Factor ANOVA

We arrange the data in a 3 x 2 table, shown in the center of the figure below. The rows show the data with regard to the first factor, the number of litters. The columns show the data with regard to the second factor, daily vitamin C supplements. We wish to test these two null hypotheses at the 0.025 significance level:

H_0: Number of litters has no effect on lifespan. (No row effect.)
H_0: Daily vitamin C supplements has no effect on lifespan. (No column effect.)

We compute a mean for each row and each column and the overall mean.

		Vitamin C supplements (in milligrams)		Mean
		50	100	Mean
Number	1	18	19	$x_{1\bullet} = 18.50$
of	5	17	19	$x_{2\bullet} = 18.00$
litters	10	10	12	$x_{3\bullet} = 11.00$
Mean		$x_{\bullet 1} = 15.00$	$x_{\bullet 2} = 16.67$	$x_{\bullet\bullet} = 15.83$

The row sum of squares is:

$$SS_r = 2\left[(18.50-15.83)^2 + (18.00-15.83)^2 + (11.00-15.83)^2\right] = 70.34$$

The column sum of squares is:

$$SS_c = 3\left[(15.00-15.83)^2 + (16.67-15.83)^2\right] = 4.20$$

The error sum of squares is:

$$SS_e = (18-18.50-15.00+15.83)^2 + (19-18.50-16.67+15.83)^2$$

$$+(17-18.00-15.00+15.83)^2 + (19-18.00-16.67+15.83)^2$$

$$+(10-11.00-15.00+15.83)^2 + (12-11.00-16.67+15.83)^2$$

$$= 0.35$$

Note that $N = (m - 1)(n - 1) = (2 - 1)(3 - 1) = 2$.

The test statistic for the hypothesis testing the row factor is:

$$F = \frac{70.34/(3-1)}{0.35/2} = 200.97 .$$

The critical value is $F_{2,2,0.025} = 39.000$. Since $200.97 > 39.000$, we conclude that the number of litters has a significant effect on the lifespan of these mammals.

The test statistic for the hypothesis testing the column factor is:

$$F = \frac{4.2/(2-1)}{0.35/2} = 24.00.$$

The critical value is $F_{1,2,0.025} = 38.506$. Since $24.00 < 38.506$, we conclude that daily vitamin C supplements has no significant effect on the lifespan of these mammals.

Answers to ?

Page	Answer
11.1	see Mean for Ungrouped Data, page 3.1 and Variance for Ungrouped Data, page 3.4
11.3	Estimating the values of and testing claims about the variance and standard deviation.

12. Correlation and Regression

Correlation

 ?   What is a scatter diagram?

Linear Correlation

Correlation measures the degree to which two data sets (samples) are related. If all data pairs, shown as points on a scatter diagram, seem to lie near a line, the correlation is called **linear**. If the y-coordinate tends to increase as the x-coordinate increases, correlation is **positive (direct).** If the y-coordinate tends to decrease as the x-coordinate increases, correlation is **negative (inverse).** If there is no relationship indicated between x and y, then there is **no correlation (uncorrelated).**

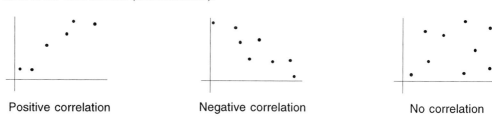

Positive correlation   Negative correlation   No correlation

Sample Correlation Coefficient

To measure correlation between the two samples, we use the **sample correlation coefficient (Pearson's Product Moment):**

$$r = \frac{n(\Sigma xy) - (\Sigma x)(\Sigma y)}{\sqrt{n(\Sigma x^2) - (\Sigma x)^2} \sqrt{n(\Sigma y^2) - (\Sigma y)^2}}$$

! r will always have a value between -1 and 1 inclusively.

Hypothesis Test for the Strength of r

To test the strength of the correlation between two samples, we use a hypothesis test with a t distribution. The testing procedure is similar to those used for mean, variance, and proportion. The population correlation coefficient is ρ. The hypotheses are H_0: $\rho = 0$ (no significant correlation) and H_1: $\rho \neq 0$ (significant correlation). The **test statistic** follows the form of previous test statistics:

$$t = \frac{r - \rho}{\sqrt{\dfrac{1 - r^2}{n - 2}}} = r\sqrt{\frac{n - 2}{1 - r^2}} \quad \text{with } df = n - 2$$

?   Why does ρ not appear in the final expression above?

Example 1

Determine the degree of correlation (that is, the correlation coefficient) between the following music and mathematics scores from ten students. Then test the hypothesis that the population correlation coefficient is 0 at $\alpha = 0.05$.

Music x	Math y	xy	x^2	y^2
80	75	6,000	6,400	5,625
70	79	5,530	4,900	6,241
82	77	6,314	6,724	5,929
83	85	7,055	6,889	7,225
86	82	7,052	7,396	6,724
87	92	8,004	7,569	8,464
85	83	7,055	7,225	6,889
84	86	7,224	7,056	7,396
93	90	8,370	8,649	8,100
90	81	7,290	8,100	6,561
840	830	69,894	70,908	69,154

$\bar{x} = 84 \quad \bar{y} = 83 \quad n = 10$

! n represents the number of pairs of data, not the total number of data from both groups.

First, find the sample correlation coefficient:

$$r = \frac{(10)(69{,}894) - (840)(830)}{\sqrt{(10)(70{,}908) - (840)^2} \ \sqrt{(10)(69{,}154) - (830)^2}} = \frac{1{,}740}{3{,}031.1} = 0.574$$

✔ r is between -1 and 1.

So, the strength of the relationship between scores measures 0.574.

Now, test the hypothesis $H_0: \rho = 0$. The alternative hypothesis is $H_1: \rho \neq 0$.

The critical values are $t_{8,0.025} = \pm 2.306$.

Using the value of r from above, the test statistic is:

$$t = r\sqrt{\frac{n-2}{1-r^2}} = 0.574\sqrt{\frac{10-2}{1-0.574^2}} = 1.983.$$

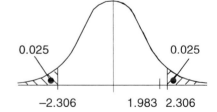

0.025   0.025

-2.306   1.983 2.306

Since $-2.306 < 1.983 < 2.306$, we fail to reject H_0.
That is, there is no significant correlation ($r = 0.574$) between music and math scores.

Determining Correlation Strength Directly

We can determine the strength of the sample correlation coefficient directly without using a
t test. A simple way to do this is to use the following guidelines.

If the value of r is ...	then we have ...
between −0.80 and −1	strong negative (inverse) correlation
between 0.80 and 1	strong positive (direct) correlation

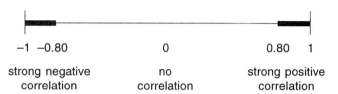

A more sophisticated way to determine the strength of the correlation is to use a **table of
critical values for the correlation coefficient**. Using this table, we can compare r to the
correct critical value and draw a conclusion about ρ without using the t test statistic.

Example 2

Using example 1 and a correlation coefficient table, we find critical values of $r = \pm 0.632$.
Since $-0.632 < 0.574 < 0.632$, there is no significant correlation between music and math
scores.

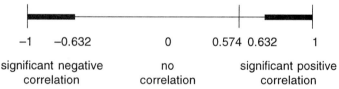

! Interpreting the strength of the correlation coefficient directly (sample r value between
the critical values means no significant correlation) is similar to interpreting the strength of
the correlation coefficient with a hypothesis test (t test value between critical values means
no significant correlation).

Observed Association Versus Causal Relationship

The correlation coefficient, r, measures the observed relationship between two data sets. It
does *not* measure what causes the relationship. Consider the next example.

Example 3

Is there a tie between the data sets in this table?

	# Whales caught worldwide	ABC Toys, Inc. common stock price ($)			
	x	y	x^2	y^2	xy
1986	29	81	841	6,561	2,349
1987	25	88	625	7,744	2,200
1988	27	85	729	7,225	2,295
1989	26	92	676	8,464	2,392
1990	24	99	576	9,801	2,376
	131	445	3,447	39,795	11,612

For these data, $r = -0.89$ indicating a highly negative (inverse) correlation between x and y. Obviously, r does not measure causation here. We cannot say that increasing stock prices of a toy company have caused a decline in the whale catch, nor vice versa.

Exercises

1. Find the correlation coefficient for the following data. Then test the hypothesis that there is no significant correlation between height and weight at $\alpha = 0.05$.

Height (in inches)	63	64	69	72
Weight (in pounds)	112	126	132	176

2. Find the correlation coefficient for the following data. Then test the hypothesis that there is no significant correlation between the two variables at $\alpha = 0.10$.

Number of doses of vitamin C	10	6	8	12
Number of days with cold	3	5	4	3

3. After inspecting the raw data in exercises 1 and 2 above, do your correlation coefficients seem reasonable? Explain your answers to 1 and 2.

Linear Regression

Slope-Intercept Form of a Line

Recall from algebra that $y = \alpha + \beta x$ is the **slope-intercept form of a line**. β is the slope $(\Delta y/\Delta x)$ and α is the y-intercept.

? Define slope and y-intercept.

Regression Line

The goal of linear regression is to find a *line* that "best fits" the scatter diagram of the sample paired data. Given a value of x, we use the linear regression equation to find the **predicted value of y**, written y'. That is,

$$y' = \alpha + \beta x$$

where
$$\alpha = \frac{(\Sigma y)(\Sigma x^2) - (\Sigma x)(\Sigma xy)}{n(\Sigma x^2) - (\Sigma x)^2}$$

$$\beta = \frac{n(\Sigma xy) - (\Sigma x)(\Sigma y)}{n(\Sigma x^2) - (\Sigma x)^2}$$

! The denominators of the formulas for α and β are the same. This saves some time when we calculate these values.

! The numerators of the formulas for β and r (from the previous section) are the same. Further, the common denominator of α and β is contained in the denominator of r. This saves more time when we calculate these values.

! Since the regression line *always* passes through $(\bar{x}, \bar{y})$, we have $\bar{y} = \alpha + \beta \bar{x}$. So, when the means are available, we can use the equation $\alpha = \bar{y} - \beta \bar{x}$ as a shortcut to find α.

Example 1

The data in the table show years of sales experience and sales volume (in millions of dollars) for eight salespersons working for an appliance manufacturer. For these data, do the following:

1) Find r
2) Test for significant correlation at $\alpha = 0.01$.
3) Find the equation of the regression line.
4) Plot the line on a scatter diagram.
5) Graphically predict the sales volume for 5 years of experience.
6) Algebraically predict the sales volume for 5 years of experience.

Years exp. x	Sales volume (millions) y	x^2	y^2	xy
1	1	1	1	1
3	2	9	4	6
4	4	16	16	16
6	4	36	16	24
8	5	64	25	40
9	7	81	49	63
11	8	121	64	88
14	9	196	81	126
56	40	524	256	364

1) $r = \dfrac{(8)(364)-(56)(40)}{\sqrt{(8)(524)-56^2}\ \sqrt{(8)(256)-40^2}} = 0.977$

2) $H_0: \rho = 0 \qquad H_1: \rho \neq 0.$
The critical values are $t_{6, 0.005} = \pm 3.707$.

The test statistic is $t = 0.977\sqrt{\dfrac{8-2}{1-0.977^2}} = 11.223$.

Since $11.223 > 3.707$, we reject H_0. That is, there is significant positive correlation between years of experience and sales volume.

3) To find the regression line, we calculate:

$\alpha = \dfrac{(40)(524)-(56)(364)}{(8)(524)-(56)^2} = 0.545$ and $\beta = \dfrac{(8)(364)-(56)(40)}{(8)(524)-(56)^2} = 0.636$

So, the regression line has the equation:
$y' = 0.545 + 0.636x$

Notice that an easier way to calculate α after finding β is to use $\bar{x} = 7$ and $\bar{y} = 5$ and the shortcut formula: $\alpha = \bar{y} - \beta\,\bar{x} = 5 - (0.636)(7) = 0.458$. This value is approximately equal to the value of α above.

4) The easiest point to graph is the y-intercept (where $x = 0$). Letting $x = 0$,
$y' = 0.545 + 0.636(0) = 0.545$.

So, $(0, 0.545)$ is a point on the graph.

To find another point of the graph, choose a convenient value for x. Suppose we pick

$x = 10$. Then, $y' = 0.545 + 0.636(10) = 6.905$.

So, $(10, 6.905)$ is also a point on the graph.

5) By the graph, when $x = 5$, it appears that $y' \approx 3.5$.

6) We check our guess algebraically with the linear regression equation:
$y' = 0.545 + 0.636(5) = 3.725$.
So, with 5 years of experience, a salesperson might expect a sales volume of $3,725,000.

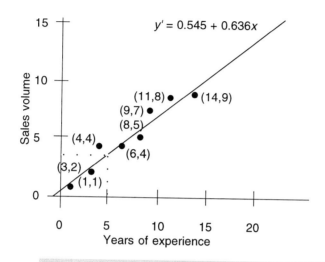

?   What is the point estimate of the mean?

Ways to Predict *y*

When r is significant, use the regression line to predict y, that is, to find y'.

When r is *not* significant, use the sample mean, $\bar{y}$, to predict y, that is, to find y'.

Example 2

Suppose the regression line is found to be $y' = -5.4 + 4.0x$, and the sample means are $\bar{x} = 2.8$ and $\bar{y} = 5.8$. Find the best predicted point estimate of y when $x = 3.0$ if:

a) the correlation test statistic is $r = 0.999$.
$r = 0.999$ shows significant positive correlation in which case we use the linear regression equation to predict y: $y' = -5.4 + 4.0(3.0) = 6.6$

b) the correlation test statistic is $r = 0.142$.
$r = 0.142$ shows no significant correlation, in which case we use the sample mean to predict y: $y' = 5.8$

Alternate Formula for *r*

This formula relates correlation to the slope of the regression line:

◆

$$r = \frac{\beta s_x}{s_y}$$

where $s_x = \sqrt{\dfrac{\Sigma (x - \bar{x})^2}{n - 1}}$ and $s_y = \sqrt{\dfrac{\Sigma (y - \bar{y})^2}{n - 1}}$

Exercises

1. Find the linear regression equation for the following data. Then use the equation to predict the weight when the height is 67.

Height (in inches)	63	64	69	72
Weight (in pounds)	112	126	132	176

2. Find the linear regression equation for the following data. Then use the equation to predict the number of days with a cold when the number of doses of vitamin C is 9.

Number of doses of vitamin C	10	6	8	12
Number of days with cold	3	5	4	3

Coefficient of Determination

In this section, we discuss how different the actual value y is from $\bar{y}$. We do this by analyzing the difference $y - \bar{y}$.

Variation

Suppose (x, y) is a point from a sample of paired observations. Consider the three values y, y', and $\bar{y}$ and this graph.

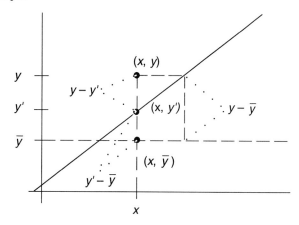

We define the following differences: $y' - \bar{y}$ is called the **explained deviation** of y from $\bar{y}$ since the difference is accountable by the regression line; $y - y'$ is called the **unexplained deviation** since we have no way of accounting for it except through sample error; $y - \bar{y}$ is the **total deviation**. So, for a given y, $y - \bar{y} = (y - y') + (y' - \bar{y})$. That is, total deviation = unexplained deviation + explained deviation.

The sums of the squares of the deviations, called **variations**, have the relationship:

$$\Sigma(y - \bar{y})^2 = \Sigma(y - y')^2 + \Sigma(y' - \bar{y})^2$$

That is, **total variation = unexplained variation + explained variation.**

! The closer the correlation coefficient r is to 1 or -1, the less variation from $\bar{y}$ there is.

?   What measure of dispersion uses a sum of squares?

Coefficient of Determination

We can interpret variation by using the **coefficient of determination**, r^2, which is derived as follows:

$$r^2 = \frac{\Sigma(y' - \bar{y})^2}{\Sigma(y - \bar{y})^2} = 1 - \frac{\Sigma(y - y')^2}{\Sigma(y - \bar{y})^2}$$

That is, $r^2 = \dfrac{\text{explained variation}}{\text{total variation}} = 1 - \dfrac{\text{unexplained variation}}{\text{total variation}}$.

The coefficient of determination has a value between 0 and 1. A value near 1 indicates that most variation in y is due to the x values (explained variation). A value near 0 indicates that most variation in y is *not* due to the x values (unexplained variation).

Example 1

Music and math scores from a sampling of students have total variation of 82.45 and unexplained variation of 18.84. Find the coefficient of determination for the paired scores. Interpret your answer.

$r^2 = 1 - \dfrac{18.84}{82.45} = 0.771$ indicates that 77.1% of the variation in math scores is due to the students' music scores. That is, 22.9% of the math scores variation is unexplained.

Correlation Coefficient: Alternative Formula

The square root of the coefficient of determination is the correlation coefficient.

✦
$$|r| = \sqrt{\dfrac{\Sigma(y' - \bar{y})^2}{\Sigma(y - \bar{y})^2}} = \sqrt{1 - \dfrac{\Sigma(y - y')^2}{\Sigma(y - \bar{y})^2}}$$

! This formula does not indicate the correct sign of the correlation coefficient. To determine whether r is positive or negative, we must inspect the graph of the regression line.

?   How does the graph of the regression line tell us the sign of r?

Example 2

Determine the strength of the relationship between the paired scores in Example 1.

The correlation coefficient is $|r| = \sqrt{0.771} = 0.878$. Note, however, that without more information, we cannot determine whether r is positive or negative.

Exercises

1. For 25 sixth graders, time spent viewing television and time spent reading per week have a total variation of 51.8 hrs and explained variation of 27.1 hrs. Find the coefficient of determination for the paired data. Interpret your answer.

2. Find the correlation coefficient of the data in exercise 1. The regression line has a negative slope.

Prediction Intervals

Prediction intervals are confidence intervals for the predicted value y'.

Standard Error of Estimate

The **standard error of estimate** measures the errors between the actual y values and the predicted y' values from the regression line.

$$s_e = \sqrt{\frac{\Sigma y^2 - \alpha \Sigma y - \beta \Sigma xy}{n-2}}$$

where α and β are from the linear regression equation

Prediction Interval for y

Similar to confidence intervals, we can determine the dependability of y', our estimate of y, by the **prediction interval for y**:

$$y' - E < y < y' + E$$

where $E = t_{n-2,\alpha/2} \cdot s_e \sqrt{1 + \frac{1}{n} + \frac{n(x_c - \bar{x})^2}{n(\Sigma x^2) - (\Sigma x)^2}}$ and x_c = given value of x

Example

Use data from Example 1 in the Linear Regression section of this chapter to find the 95% prediction interval of y when $x_c = 5$.

Using the linear regression equation, when $x_c = 5$,
$y' = 0.545 + 0.636x = 0.545 + 0.636(5) = 3.72$.

So, $s_e = \sqrt{\frac{256 - (0.545)(40) - (0.636)(364)}{6}} = 0.67$, $t_{6,0.025} = 2.447$,

and $E = (2.447)(0.67)\sqrt{1 + \frac{1}{8} + \frac{8(5-7)^2}{8(524) - 56^2}} = 1.76$.

Thus,
$$y' - E < y < y' + E$$
$$3.72 - 1.76 < y < 3.72 + 1.76$$
$$1.96 < y < 5.48$$

Exercise

Find the 98% prediction interval of y if $x_c = 5$ for these data.

Number of children in family	1	2	3	4
Amount of garbage recycled weekly (lbs)	24	26	31	35

Solutions to Exercises

Correlation

1. $r = \dfrac{(4)(36,900) - (268)(546)}{\sqrt{(4)(18,010) - (268)^2}\,\sqrt{(4)(76,820) - (546)^2}} = \dfrac{1,272}{1406.920} = 0.904$

    $H_0: \rho = 0$ and $H_1: \rho \neq 0$.

    The critical values are $t_{2,0.025} = \pm 4.303$.

    Using the value of r from above, the test statistic is:

    $t = r\sqrt{\dfrac{n-2}{1-r^2}} = 0.904\sqrt{\dfrac{4-2}{1-0.904^2}} = 2.990$

    Since $-4.303 < 2.990 < 4.303$, we fail to reject H_0. That is, there is no significant correlation between height and weight in this sample.

2. $r = \dfrac{(4)(128) - (36)(15)}{\sqrt{(4)(344) - (36)^2}\,\sqrt{(4)(59) - (15)^2}} = \dfrac{-28}{29.665} = -0.944$

    $H_0: \rho = 0$ and $H_1: \rho \neq 0$.

    The critical values are $t_{2,0.05} = \pm 2.920$.

    Using the value of r from above, the test statistic is:

    $t = r\sqrt{\dfrac{n-2}{1-r^2}} = -0.944\sqrt{\dfrac{4-2}{1-(-0.944)^2}} = -4.046$

    Since $-4.046 < -2.920$, we reject H_0. That is, there is significant negative correlation between vitamin C dosage and length of illness in this sample.

3. In exercise 1, the data indicate a strong positive correlation. However, with only four pairs of data, we cannot be sure of significant correlation. Note the dissimilar weight in the fourth pair of data. In exercise 2, the data indicate a strong negative correlation. Its significance is supported by the hypothesis test.

Linear Regression

1. Let x = height and y = weight. To find the regression line, we calculate:

    $\alpha = \dfrac{(546)(18,010) - (268)(36,900)}{(4)(18,010) - (268)^2} = -258.06$ and $\beta = \dfrac{(4)(36,900) - (268)(546)}{(4)(18,010) - (268)^2} = 5.89$

    So, the regression line has the equation:
    $y' = -258.06 + 5.89x$

    When $x = 67$, $y' = -258.06 + 5.89(67) = 136.57$.

2. Let x = dosage and y = length of illness. To find the regression line, we calculate:

$$\alpha = \frac{(15)(344)-(36)(128)}{(4)(344)-(36)^2} = 6.9 \text{ and } \beta = \frac{(4)(128)-(36)(15)}{(4)(344)-(36)^2} = -0.35$$

So, the regression line has the equation:
$y' = 6.9 - 0.35x$

When $x = 9$, $y' = 6.9 - 0.35(9) = 3.75$.

Coefficient of Determination

1. $r^2 = \dfrac{27.1}{51.8} = 0.523$ indicates that 52.3% of the variation in reading time is due to TV

    viewing time. That is, 47.7% of the reading time variation is unexplained.

2. $|r| = \sqrt{0.523} = 0.723$. Since the regression line has a negative slope, $r = -0.723$.

Prediction Intervals

$\Sigma x = 10$, $\Sigma y = 116$, $\Sigma x^2 = 30$, $\Sigma y^2 = 3438$, $\Sigma xy = 309$, and $n = 4$. The correlation coefficient is
$r = 0.99$. The regression line is $y' = 19.5 + 3.8x$. When $x_c = 5$, $y' = 19.5 + 3.8(5) = 38.5$

So, $s_e = \sqrt{\dfrac{3438-(19.5)(116)-(3.8)(309)}{2}} = 0.95$

and $E = (6.965)(0.95)\sqrt{1+\dfrac{1}{4}+\dfrac{4(5-2.5)^2}{4(30)-10^2}} = 10.5$

Thus, the 98% prediction interval for y is
$$y' - E < y < y' + E$$
$$38.5 - 10.5 < y < 38.5 + 10.5$$
$$28 < y < 49$$

Answers to ?

Page	Answer
12.1	see Scatter Diagram: An Example, page 2.12
12.1	Since H_0: $\rho = 0$, we can drop it from the formula.
12.5	Slope measures the tilt or angle of the line. The y-intercept is the point where the line crosses the y-axis.
12.7	the sample mean
12.9	variance
12.10	Positive slope indicates positive correlation; negative slope indicates negative correlation.

13. Goodness of Fit Tests and Contingency Tables

Goodness of Fit Tests

In this section, we extend *binomial* experiments to *multinomial* experiments. That is, instead of an experiment having only two outcomes, it can have several. We test the *observed* sample data against their *expected* values in a χ^2 hypothesis test. This kind of hypothesis test is called a **goodness of fit test**.

Expected Frequency

The **expected frequency** of an outcome of a multinomial experiment is

✦
$$E = n \cdot p$$

where n is the number of trials and p is the favorable probability.

? What other concept uses this same formula?

Hypotheses

The null hypothesis of the χ^2 test states that there is no significant difference between the observed and expected values. The alternative hypothesis states that there is a significant difference.

! Since χ^2 is always positive, we *always* have a right-tailed test.

Test Statistic

The **test statistic** for a goodness of fit test is from the chi-square distribution:

✦
$$\chi^2 = \Sigma \frac{(O-E)^2}{E}$$

where O is the observed frequency of an outcome
E is the expected frequency of an outcome

The **critical value** has $df = m - 1$, where m is the number of classes or categories of data.

! Use this test statistic if and only if *each E* value ≥ 5.

Example

A die is rolled 150 times with the following results. Test whether the die is perfectly balanced at $\alpha = 0.05$.

Value on die	1	2	3	4	5	6
Frequency	26	23	33	34	18	16

The hypotheses are:

H_0: $p_1 = \ldots = p_6 = 1/6$

H_1: At least one of these probabilities is not equal to the others

Tabulate the data as follows to calculate the test statistic.

Class (category) m	Observed freq. O	Prob. p	Expected freq. $E = np$	$O - E$	$(O - E)^2$	$\dfrac{(O - E)^2}{E}$
1	26	1/6	25	1	1	0.04
2	23	1/6	25	-2	4	0.16
3	33	1/6	25	8	64	2.56
4	34	1/6	25	9	81	3.24
5	18	1/6	25	-7	49	1.96
6	16	1/6	25	-9	81	3.24
	150	1.00	150	0		test $\chi^2 = 11.20$

✔ These two columns total n.

✔ This column totals 1.00.

✔ This column totals 0.

$\alpha = 0.05$, $df = m - 1 = 5$, and a right-tailed test yield a critical value of $\chi^2 = 11.071$. Since $11.20 > 11.071$, we reject H_0 that the die is perfectly balanced.

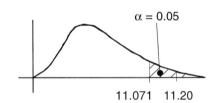

$\alpha = 0.05$

11.071   11.20

?   What are the differences in the fifth column of the table above called?

Exercise

The education levels of the adult population of a midwestern state, along with the respective probabilities of randomly selecting a state resident whose education concluded at that level, are shown in the table. A town in that state believes that its adult population has the same breakdown in education levels as the state. A sample of 100 citizens from the town are polled for education level. The results are also shown in the table. At $\alpha = 0.05$, test the claim that the town has the same education-level breakdown as the state.

Highest education level gained	State probabilities	Town's head count
8th grade	0.05	3
12th grade	0.60	50
College	0.30	40
Graduate school	0.05	7

Contingency Tables

Contingency tables are used in deciding whether one variable is independent of another. Categories of the two variables are arranged in rows and columns of a table. If there are r categories of the first variable and c categories of the second variable, we arrange the data in an $r \times c$ contingency table. We will use the χ^2 distribution as in the previous section.

Expected Frequency

Expected frequency for a particular row and column in the contingency table is:

◆
$$E = \frac{(\text{row total})(\text{column total})}{\text{grand total}}$$

Hypotheses

H_0 states that the two variables are independent. H_1 states that there is some dependence between the variables.

Test Statistic

The **test statistic** for a contingency table is:

◆
$$\chi^2 = \Sigma \frac{(O-E)^2}{E}$$

The **critical value** has $df = $ (number of rows $- 1$)(number of columns $- 1$).

where O is the observed frequency of an outcome
E is the expected frequency of an outcome

❗ Each E must be greater than or equal to 5 and the test is *always* right-tailed.

Example

We wish to decide whether a person's complexion is related to a preference for certain brands of suntan lotions. Three brands of lotion, A, B, and C, are offered to 100 sunbathers categorized in three complexion categories: fair, medium, dark. The results are classified into a 3x3 contingency table as shown. At $\alpha = 0.01$, test the claim that complexion and suntan lotion are independent.

		Complexion		
		Fair	Medium	Dark
Suntan lotion	A	10	25	15
	B	15	10	5
	C	10	5	5

First, we add totals and expected frequencies to the contingency table. Expected frequencies are shown in parentheses after observed frequencies. For example, the expected value of fair complexioned sunbathers preferring brand A is $E = \dfrac{(50)(35)}{100} = 175$.

		Complexion			Total
		Fair	Medium	Dark	
Suntan lotion	A	10 (17.5)	25 (20)	15 (12.5)	50
	B	15 (10.5)	10 (12)	5 (7.5)	30
	C	10 (7)	5 (8)	5 (5)	20
	Total	35	40	25	100

The sum of the row totals equals the sum of the column totals.

Next, $df = $ (number of rows − 1)(number of columns − 1) $= (3 − 1)(3 − 1) = 4$

So, the critical value at $\alpha = 0.01$ is $\chi^2 = 13.277$.

The test statistic is:

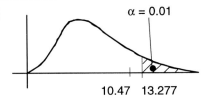

$\alpha = 0.01$

10.47   13.277

$$\chi^2 = \frac{(10-17.5)^2}{17.5} + \frac{(25-20)^2}{20} + \frac{(15-12.5)^2}{12.5} + ... + \frac{(5-5)^2}{5} = 10.47.$$

Since $10.47 < 13.277$, we do not reject H_0. That is, complexion and suntan lotion are independent.

Exercise

The education and income class of 1000 people of a midwestern town are shown in the table. Test the claim that the town's education and income levels are independent. Use $\alpha = 0.05$.

Education level:	Income class		
	Lower	Middle	Upper
8th grade	20	50	0
12th grade	60	150	40
College	80	270	60
Graduate school	70	100	100

Solutions to Exercises

Goodness of Fit Tests

H_0: The state probabilities hold for the town's population.
H_1: At least one of the town's probabilities is not equal to the corresponding state probability.

Class (category)	Observed freq. O	Prob. p	Expected freq. $E = np$	$O - E$	$(O - E)^2$	$\dfrac{(O - E)^2}{E}$
8th grade	3	0.05	5	−2	4	0.80
12th grade	50	0.60	60	−10	100	1.67
College	40	0.30	30	10	100	3.33
Graduate	7	0.05	5	2	4	0.80
	100	1.00	100	0		test $\chi^2 = 6.60$

$\alpha = 0.05$, $df = m - 1 = 3$, and a right-tailed test yield a critical value of 7.815. Since $6.60 < 7.815$, we fail to reject H_0. The town has the same education level breakdown as the state.

Contingency Tables

Education level:	Income class			Total
	Lower	Middle	Upper	
8th grade	20 (16.1)	50 (39.9)	0 (14.0)	70
12th grade	60 (57.5)	150 (142.5)	40 (50.0)	250
College	80 (94.3)	270 (233.7)	60 (82.0)	410
Graduate school	70 (62.1)	100 (153.9)	100 (54.0)	270
Total	230	570	200	1000

Next, $df = $ (number of rows − 1)(number of columns − 1) $= (4 - 1)(3 - 1) = 6$. So, the critical value at $\alpha = 0.05$ is $\chi^2 = 12.592$.

The test statistic is

$$\chi^2 = \frac{(20-16.1)^2}{16.1} + \frac{(50-39.9)^2}{39.9} + \frac{(0-14.0)^2}{14.0} + \cdots + \frac{(100-54.0)^2}{54.0} = 92.78.$$

Since $92.78 < 12.592$, we do not reject H_0. There is some dependence between education and income.

Answers to ?

Page	Answer
13.1	expected value
13.1	deviations

Quick Reference

Quick Reference Contents

Page numbers in parentheses indicate where the topic is explained in greater detail.

2. Describing Data

3. Summarizing Data

4. Probability and Counting

5. Discrete Random Variables

6. Normal Random Variables

7. Distributions of Sampling Statistics

8. Estimation

9. Testing Hypotheses

10. Hypothesis Tests for Two Populations

11. Analysis of Variance (ANOVA)

12. Correlation and Regression

13. Goodness of Fit Tests and Contingency Tables

2. Describing Data

Which Distribution to Use (pp. 2.2–2.4, 2.9–2.10)

Use this table for suggestions on how to summarize your data.

If you want to show ...	and your data is ...	then use a ...
data and frequency	ungrouped	frequency distribution
	grouped	frequency distribution or stem-and-leaf plot
data and cumulative frequency	ungrouped or grouped	cumulative frequency distribution
data and relative frequency	ungrouped or grouped	relative frequency distribution
data and cumulative relative frequency	ungrouped or grouped	cumulative relative frequency distribution

Which Graph to Use (p. 2.6–2.13)

Use this table for suggestions on how to graph your data.

If you want to show ...	and your data is ...	then use a ...
data and frequency	ungrouped	bar chart or frequency polygon
	grouped	histogram or frequency polygon
data and cumulative frequency	ungrouped or grouped	ogive
data and relative frequency	ungrouped	pie chart
	grouped	relative histogram or relative frequency polygon
data and cumulative relative frequency	ungrouped or grouped	ogive
data and dispersion	ungrouped	boxplot
data and relationships	paired	scatter diagram

Grouping Data into Class Intervals (pp. 2.1–2.2)

To group raw data into class intervals, follow these steps.

Step	Action
1	Arrange the data from least to greatest.
2	Subjectively determine the number of classes that display the most information about your data (generally between 5 and 20 classes).
3	Determine the class width by using: $$\text{class width} = \frac{\text{range}}{\text{number of classes}}, \text{ rounded up}$$ where range = highest value – lowest value.
4	Calculate the lower and upper class limit for each interval starting with a number less than or equal to the lowest value. If the highest value is not included in the last interval, choose a different starting point and recalculate.

Frequency Distribution (p. 2.3)

A frequency distribution is a two-column table. It is the basis for other distributions. To construct a frequency distribution for ungrouped or grouped data, follow these steps.

Step	Action
1	In the data (first) column, enter your raw data or class intervals
2	In the frequency (second) column, for each entry in the data column, tally the number of times each value occurs (n). The sum of this column equals the total number of values (N)

Cumulative Frequency Distribution (p. 2.3)

This is a three-column table. It accumulates the frequencies through the class intervals. To construct a cumulative frequency distribution, follow these steps.

Step	Action
1	Construct a frequency distribution for the data
2	For each row in the table, add the frequency (n) to those of all preceding rows, and place the sum in a cumulative frequency (third) column. The last entry in this column equals the total number of values (N),

Relative Frequency Distribution (p. 2.3)

This table has three columns. It shows the percentage of values in each class. To construct a relative frequency distribution, follow these steps.

Step	Action
1	Construct a frequency distribution for the data.
2	For each row in the table, calculate the relative frequency (n/N) in a relative frequency (third) column. The sum of this column equals 1 (100%).

Cumulative Relative Frequency Distribution (p. 2.4)

There are four columns in this table. It accumulates the percentages from the relative frequency distribution. To construct a cumulative relative frequency distribution, follow these steps.

Step	Action
1	Construct a *relative* frequency distribution for the data.
2	For each row in the table, add the relative frequency to those of all preceding rows and place the sum in a cumulative relative frequency (fourth) column. The last entry in this column equals 1 (100%).

Pie Chart (p. 2.6)

A pie chart shows the relative size of grouped data as sectors of a circle. To construct a pie chart, follow these steps.

Step	Action
1	Construct a relative frequency distribution for the data.
2	Divide the pie into sectors that correspond in size to the relative frequencies: number of degrees in the sector = relative frequency times 360.
3	Label each sector with the data class (or name) and the relative frequency as a percentage.

Bar Chart (p. 2.6)

Use a bar chart with ungrouped data to show frequencies. To construct a bar chart, follow these steps.

Step	Action
1	Construct a frequency distribution for the ungrouped data.
2	Plot the ungrouped data along the horizontal axis.
3	Plot the frequencies along the vertical axis.
4	For each data value (horizontal axis), draw a bar to the height of the frequency (vertical axis) for that value.

Histogram or Relative Histogram (pp. 2.6–2.7)

Use histograms to show frequencies of grouped data. To construct a histogram or relative histogram, follow these steps.

Step	Action
1	Construct a frequency distribution (for a histogram) or a relative frequency distribution (for a relative histogram).
2	Plot the class boundaries along the horizontal axis. Start with the smallest class boundary at the origin.
3	Plot the frequencies (for a histogram) or the relative frequencies (for a relative histogram) along the vertical axis.
4	For each class interval (horizontal axis), draw a bar to the height of the frequency (vertical axis) for that interval.

Frequency Polygon (p. 2.7)

A frequency polygon is another way to display grouped data and their frequencies. To construct a frequency polygon, follow these steps.

Step	Action
1	Construct a frequency distribution for the data.
2	Plot the class marks along the horizontal axis.
3	Plot the frequencies along the vertical axis.
4	Plot a point for each class mark–frequency pair.
5	Starting at the origin, connect the points. End the polygon on the horizontal axis.

Ogive (p. 2.8)

An ogive is also known as a cumulative frequency polygon. It shows the grouped data and the accumulated frequencies. To construct an ogive, follow these steps.

Step	Action
1	Construct a cumulative frequency distribution for the data.
2	Plot the class boundaries along the horizontal axis.
3	Plot the cumulative frequencies along the vertical axis.
4	Plot a point for each class boundary–frequency pair.
5	Starting at the origin, connect the points. The last point has a cumulative frequency of 1.

Stem-and-Leaf Plot (p. 2.9)

A stem-and-leaf plot preserves the raw data and also gives a picture of the data. This is a two-column table. To construct a stem-and-leaf plot, follow these steps.

Step	Action
1	Arrange the data from least to greatest.
2	Subjectively determine the stems (leftmost digit(s) of the data) that display the most information about your data.
3	Enter the stems in the first column.
4	Enter the leaves (remaining rightmost digit(s)) in the second column. Turned sideways, this column has the same profile as a histogram of the data.

Boxplot (pp. 2.10–2.11)

A boxplot can be used to identify improperly recorded data. To construct a boxplot, follow these steps.

Step	Action
1	Arrange the data from least to greatest.
2	Identify the 5-number summary: minimum value, lower hinge, median, upper hinge, and maximum value.
3	Plot the 5-number summary on a number line.
4	Draw a box from the lower hinge to the upper hinge including the median.

3. Summarizing Data

Mean of a Sample (p. 3.1)

The mean is the most common and useful averaging technique.

For ungrouped data,

$$\bar{x} = \frac{\Sigma x}{n}$$

where x is datum,
n is the number of data

For grouped data,

$$\bar{x} = \frac{\Sigma fx}{\Sigma f} = \frac{\Sigma fx}{n}$$

where x is the class midpoint,
f is the class frequency

Median of a Sample (p. 3.2)

The median is the value literally in the middle of the other data.

Step	Action
1	Arrange the data by magnitude.
2	**If the number of data is:** **then the median is:** odd the middle value. even the mean of the two middle values.

Mode of a Sample (p. 3.2)

The mode is the value that occurs most often.

Step	Action
1	Arrange the data by magnitude.
2	The mode is the value occurring most often and more than once. If no value occurs more than once, there is no mode. There can be more than one mode.

Midrange of a Sample (p. 3.2)

The midrange is the mean of the first and last data values.

Weighted Mean of a Sample (p. 3.3)

$$\text{Weighted Mean} = \frac{\Sigma wx}{\Sigma w}, \text{ where } x \text{ is datum with corresponding weight } w.$$

Variance of a Sample (pp. 3.4–3.5)

Variance is the most common and useful way to measure disperson of data around the mean.

For ungrouped data,

$$s^2 = \frac{\Sigma(x-\bar{x})^2}{n-1} = \frac{n(\Sigma x^2) - (\Sigma x)^2}{n(n-1)}$$

where x is datum,
n is the number of data

For grouped data,

$$s^2 = \frac{\Sigma fx^2 - n\bar{x}^2}{n-1} = \frac{n\Sigma(fx^2) - (\Sigma fx)^2}{n(n-1)}$$

where x is the class midpoint,
f is the class frequency

Standard Deviation of a Sample (p. 3.6)

standard deviation $= \sqrt{\text{variance}}$, that is, $s = \sqrt{s^2}$

Standard Score, or z Score (p. 3.7)

Convert a score x to a standard score z by $z = \dfrac{x - \bar{x}}{s}$, where $\bar{x}$ and s are from the sample data.

Percentile, Decile, or Quartile of a Given Value (pp. 3.7–3.8)

Given a value x,
quartile of x = (# of data less than x / total number of data) · 4, rounded off
decile of x = (# of data less than x / total number of data) · 10, rounded off
percentile of x = (# of data less than x / total number of data) · 100, rounded off

Value of a Given Percentile, Decile, or Quartile (p. 3.9)

Given a quartile Q_k, decile D_k, or percentile P_k, find its value by using this equation to solve this type of problem:

Step	Action		
1	Arrange the data in ascending order.		
2	Find $V = (k / l) \cdot n$	where	$l = 4$ for a quartile, 10 for a decile, 100 for a percentile
		and	n = number of data
3	**If V is:**		**then the value of $Q_k, D_k,$ or P_k is the:**
	a whole number		mean of the Vth datum and V+1st datum.
	not a whole number, *round up V*;		Vth datum.

4. Probability and Counting

Probability Basics (p. 4.1)

The probability of event A is $P(A) = \dfrac{m}{n}$, where n is the total number of equally likely outcomes of the experiment, and m is the number of outcomes that pertain to event A.

Number of Outcomes (p. 4.2)

When an experiment can be viewed as a series of trials of a simple experiment, the number of outcomes of the overall experiment $= a^b$, where a is the number of outcomes per trial and b is the number of trials in the overall experiment.

Addition Rule (p. 4.4)

If events A and B are disjoint, $P(A \text{ or } B) = P(A \cup B) = P(A) + P(B)$.
If events A and B are not disjoint, $P(A \text{ or } B) = P(A \cup B) = P(A) + P(B) - P(A \text{ and } B)$.

Multiplication Rule (pp. 4.5–4.6)

If events A and B are independent, $P(A \text{ and } B) = P(A \cap B) = P(A) \cdot P(B)$.
If events A and B are not independent, $P(A \text{ and } B) = P(A \cap B) = P(A) \cdot P(B|A)$.

Bayes' Theorem (p. 4.7)

When the probability of the outcome of an event A is influenced by another event B:

$$P(A|B) = \frac{P(B|A) \cdot P(A)}{P(B|A) \cdot P(A) + P(B|A^c) \cdot P(A^c)}$$

Methods of Counting (pp. 4.8–4.9)

The Fundamental Rule of Counting says that if one action can be performed in m ways, and another action can be performed in n ways, then both actions can be performed in $m \cdot n$ ways.

n factorial is defined as $n! = n(n-1)(n-2) \cdot \ldots \cdot 2 \cdot 1$. There is one exception to this rule: $0! = 1$

Use combinations when the order in which the items are selected is *not* important.

$$_nC_i = \frac{n!}{i!(n-i)!}$$

Use permutations when the order in which the items are selected *is* important.

$$_nP_i = \frac{n!}{(n-i)!}$$

In both of the formulas above, $n =$ total number of items, and $i =$ number of items pertaining to event A.

5. Discrete Random Variables

Verifying a Probability Distribution (p. 5.1)

A probability distribution must meet two criteria:

1) $0 \leq P(x) \leq 1$, for each x and   2) $\Sigma P(x) = 1$

Describing a Probability Distribution (pp. 5.3–5.4)

Describe any probability distribution with the expected value, variance, and standard deviation:

$$E(x) = \mu = \Sigma x \cdot P(x)$$

$$Var(x) = \sigma^2 = \left[\Sigma x^2 \cdot P(x)\right] - \mu^2 = E(x^2) - \mu^2$$

and $SD(x) = \sigma = \sqrt{Var(x)}$

Properties of the expected value and variance are:

$E(cx) = cE(x)$	(1)	$Var(cx) = c^2 Var(x)$	(4)
$E(x+c) = E(x) + c$	(2)	$Var(x+c) = Var(x)$	(5)
$E(x+y) = E(x) + E(y)$	(3)	$Var(x+y) = Var(x) + Var(y)$	(6) *

where x and y are random variables, and c is a constant.

* Equation (6) holds only when x and y are independent, that is, when the value of one does not influence the probability of the other.

Binomial Probability with the Binomial Theorem (pp. 5.6–5.7)

The probability that event A occurs i out of n times is:

$P(A \text{ occurs } i \text{ times}) = {}_nC_i \cdot p^i \cdot q^{n-i}$ where n = number of trials
i = number of successes in n trials
$p = P(A)$ = probability of success
$q = 1 - p$ = probability of failure

$${}_nC_i = \frac{n!}{i!(n-i)!}$$

Binomial Probability Distribution (p. 5.9)

Describe a *binomial* probability distribution with the expected value, variance, and standard deviation:

$E(x) = np$, $Var(x) = npq$, and $SD(x) = \sqrt{Var(x)}$ where n = number of trials
p = favorable probability
q = unfavorable probability

6. Normal Random Variables

Standard Normal Probability Distribution (p. 6.2)

Given a standard normal z score, find its corresponding probability by following these steps.

Step	Action
1	Sketch the standard normal curve and label it with $\mu = 0$ and with values from your problem.
2	Shade the region under the curve where the probability is to be determined.
3	Find the probability in a standard normal probabilities table.

Normal Probabilities (pp. 6.5–6.6)

Given a normal random variable, x, find its corresponding probability by following these steps.

Step	Action
1	Convert the random variable, x, to a z score by using $z = \dfrac{x - \mu}{\sigma}$.
2	Sketch the standard normal curve and label it with the given values and z scores from your problem.
3	Shade the region under the curve where the probability is to be determined.
4	Find the probabilities of the z scores in a standard normal probabilities table.
5	If necessary, add or subtract the probabilities to correspond to the shaded region under the curve.

If the problem involves two independent random variables, x and y, use these formulas to standardize before finding the probability:

$$E(x + y) = E(x) + E(y) = \mu_x + \mu_y \qquad \text{and} \qquad SD(x + y) = \sqrt{\sigma_x^2 + \sigma_y^2}$$

Percentiles of Standard Normal Random Variables (p. 6.7)

To find the value, z_α, of a percentile in a *standard* normal distribution, follow these steps.

Step	Action
1	To determine the percentile associated with z_α, calculate $100(1 - \alpha)$. This tells what percentage of values are to the left of z_α.
2	Sketch the normal curve and label it with values from your problem.
3	**If z_α is on this side of the normal curve:** **then find the closest value to:**
	right side $100(1 - \alpha)$ in the table.
	left side $100(\alpha)$ in the table.
4	Read the corresponding z score. This is the *value* of the percentile. **If z_α is on this side of the normal curve:** **then z_α is equal to:**
	right side the value in the table.
	left side the negative of the value in the table.

Percentiles of Normal Random Variables (p. 6.8)

To find the value of the percentile, z_α, in a normal distribution, follow these steps.

Step	Action
1	Follow the steps for Percentiles of Standard Normal Random Variables listed above.
2	Convert the z score to a normal random variable, x, by using $x = \mu + z\sigma$.

7. Distributions of Sampling Statistics

Parameters from the Central Limit Theorem (p. 7.1)

This table summarizes the parameters from the Central Limit Theorem.

	If the population distribution has:	then the sampling distribution of $\bar{x}$ has:
Random variable	x	$\bar{x}$
Mean	μ	$\mu_{\bar{x}} = \mu$
Standard deviation	σ	$\sigma_{\bar{x}} = \sigma/\sqrt{n}$

Summary for Standard Error of the Mean Formulas (p. 7.3)

This table summarizes which standard error formula to use depending on how you sample.

If you are sampling:	then use:
with replacement	$\sigma_{\bar{x}} = \dfrac{\sigma}{\sqrt{n}}$
without replacement and $n \le 0.05N$ (small sample)	$\sigma_{\bar{x}} = \dfrac{\sigma}{\sqrt{n}}$
without replacement and $n > 0.05N$ (large sample)	$\sigma_{\bar{x}} = \dfrac{\sigma}{\sqrt{n}} \sqrt{\dfrac{N-n}{N-1}}$

Normal as Approximation to the Binomial Distribution (pp. 7.5–7.6)

To use the normal approximation to the binomial distribution, follow these steps.

Step	Action
1	Determine if *both* $np \ge 5$ *and* $nq \ge 5$. If either inequality does not hold, you cannot use the normal approximation. Do not proceed.
2	Use the correction for continuity to expand the region under the curve to either side.
3	Determine the probability by standardizing the random variable.

8. Estimation

Interval Estimators of the Population Mean (pp. 8.1–8.4)

This table helps you decide whether to use a z distribution or a t distribution when estimating the population mean.

Use the z distribution when *either* of these conditions is met:	Use the t distribution when *all* of these conditions are met:
• σ is known and n is any size, **or** • The sample size is $n > 30$ and σ is unknown (but estimated by s).	• The sample is small ($n \leq 30$), **and** • σ is unknown, **and** • The population is essentially normal.

Follow these steps to construct a $1 - \alpha$ confidence interval for the population mean.

Step	Action	z distribution	t distribution:
1	Find the critical value.	$z_{\alpha/2}$	$t_{df,\alpha/2}$
2	Find the maximum error of the mean.	$E = z_{\alpha/2} \dfrac{\sigma}{\sqrt{n}}$ If σ is unknown, use s as an estimate.	$E = t_{df,\alpha/2} \dfrac{s}{\sqrt{n}}$
3	Construct the confidence interval.	$\bar{x} - E < \mu < \bar{x} + E$	

To find the sample size necessary to ensure a level of confidence based on a maximum error, E, of the mean, use $n = \left(\dfrac{z_{\alpha/2}\,\sigma}{E} \right)^2$.

Interval Estimators of the Population Variance (p. 8.6)

Follow these steps to construct a $1 - \alpha$ confidence interval for the population variance, σ^2, and the population standard deviation, σ. Note that this procedure uses a χ^2 distribution.

Step	Action
1	Find the critical values, χ_L^2 and χ_R^2.
2	Construct the confidence intervals: $$\frac{(n-1)s^2}{\chi_R^2} < \sigma^2 < \frac{(n-1)s^2}{\chi_L^2} \quad \text{and} \quad \sqrt{\frac{(n-1)s^2}{\chi_R^2}} < \sigma < \sqrt{\frac{(n-1)s^2}{\chi_L^2}}$$

Interval Estimators of the Population Proportion (pp. 8.8–8.9)

Follow these steps to construct a $1 - \alpha$ confidence interval for the population proportion, p. Note that this procedure uses a z distribution.

Step	Action
1	Find the critical value, $z_{\alpha/2}$.
2	Find $\hat{p} = x/n$ and $\hat{q} = 1 - \hat{p}$.
3	Use these results to find the maximum error of the estimate of p, $E = z_{\alpha/2}\sqrt{\dfrac{\hat{p}\hat{q}}{n}}$.
4	Construct the confidence interval, $\hat{p} - E < p < \hat{p} + E$.

To find the sample size necessary to ensure a maximum error, E, of the population proportion, p, use the formula $n = z_{\alpha/2}^2 \dfrac{\hat{p}\hat{q}}{E^2}$.

9. Testing Hypotheses

Performing a Hypothesis Test (p. 9.1)

Follow these steps when testing a hypothesis concerning the population mean, proportion, variance, or standard deviation.

Step	Action
1	Formulate the null and alternative hypotheses based on the wording of the problem.
2	Sketch and label the curve with available information.
3	Determine the critical value(s) using the significance level and include it (them) on the sketch.
4	Calculate the test statistic and mark it on the sketch.
5	Compare the test statistic to the critical value(s). With the help of the sketch, draw a conclusion about the null hypothesis.

Stating the Hypotheses (p. 9.1)

Relational symbols are paired in H_0 and H_1 as follows.

Symbols are correctly used if:	Example:	
H_0 contains $=$ and H_1 contains $\neq$	H_0: $\mu = 100$	H_1: $\mu \neq 100$
H_0 contains $\geq$ and H_1 contains $<$	H_0: $\mu \geq 100$	H_1: $\mu < 100$
H_0 contains $\leq$ and H_1 contains $>$	H_0: $\mu \leq 100$	H_1: $\mu > 100$

Sketching the Curve of a Hypothesis Test (p. 9.3)

Use these guidelines to sketch the distribution for the test you are performing.

If H_1 contains:	the test is:	and:
$\neq$	two-tailed	α is divided evenly into each tail.
$<$	left-tailed	α is completely in the left tail.
$>$	right-tailed	α is completely in the right tail.

Distributions for Different Hypothesis Tests (pp. 9.4, 9.8, 9.9, 9.11)

The type of hypothesis test you perform determines which distribution to use.

If you are testing the:	when:	then, use the	with test statistic:
population mean, μ	σ is known and the population is normal	z distribution	$z = \dfrac{\bar{x} - \mu}{\sigma/\sqrt{n}}$
population mean, μ	$n > 30$ and the population is normal (If σ is unknown, use s.)	z distribution	$z = \dfrac{\bar{x} - \mu}{\sigma/\sqrt{n}}$
population mean, μ	σ is unknown, $n \leq 30$, and the population is normal	t distribution	$t = \dfrac{\bar{x} - \mu}{s/\sqrt{n}}$
population proportion, p	$np \geq 5$ and $nq \geq 5$	z distribution	$z = \dfrac{\hat{p} - p}{\sqrt{pq/n}}$
population variance, σ^2, or standard deviation, σ	the population is normal	χ^2 distribution	$\chi^2 = \dfrac{(n-1)s^2}{\sigma^2}$

10. Hypothesis Tests for Two Populations

Five Types of Hypothesis Tests for Two Means (pp. 10.2, 10.4, 10.7, 10.11–10.12)

There is only one hypothesis test for two population variances and only one test for two population proportions. However, there are *five* kinds of hypothesis tests for two population means. Here is a summary of these five test:

Samples	Population variances	Sample sizes	Assumptions about population variances	Distribution
dependent				t
independent	known			z
independent	unknown	large		z
independent	unknown	small	assumed equal	F and t
independent	unknown	small	assumed unequal	F and t

To determine which test to use, answer these questions about the problem.

> Are the samples dependent or independent?

If samples are dependent, then:
the test statistic is

$$t = \frac{\bar{d} - \mu_d}{s_d / \sqrt{n}}$$

and the maximum error for the confidence interval is

$$E = t_{df, \alpha/2} \frac{s_d}{\sqrt{n}} \text{ with } df = n - 1$$

> If samples are independent, are the population variances known or unknown?

If σ_1^2 and σ_2^2 are known, then:
the test statistic is

$$z = \frac{\bar{x}_1 - \bar{x}_2}{\sqrt{\frac{\sigma_1^2}{n_1} + \frac{\sigma_2^2}{n_2}}}$$

and the maximum error for the confidence interval is

$$E = z_{\alpha/2} \sqrt{\frac{\sigma_1^2}{n_1} + \frac{\sigma_2^2}{n_2}}$$

> If the population variances are unknown, are the sample sizes *both* greater than 30?

If *both* $n_1 > 30$ and $n_2 > 30$, then:
the test statistic is

$$z = \frac{\bar{x}_1 - \bar{x}_2}{\sqrt{\frac{s_1^2}{n_1} + \frac{s_2^2}{n_2}}}$$

and the maximum error for the confidence interval is

$$E = z_{\alpha/2} \sqrt{\frac{s_1^2}{n_1} + \frac{s_2^2}{n_2}}$$

If *either* n_1 or $n_2 \le 30$, then, by using the F test, can we assume $\sigma_1^2 = \sigma_2^2$?

If we can assume that $\sigma_1^2 = \sigma_2^2$, then:
the test statistic is

$$t = \frac{\bar{x}_1 - \bar{x}_2}{\sqrt{\dfrac{(n_1 - 1)s_1^2 + (n_2 - 1)s_2^2}{n_1 + n_2 - 2}\left(\dfrac{1}{n_1} + \dfrac{1}{n_2}\right)}}$$

and the maximum error for the confidence interval is

$$E = t_{df, \alpha/2} \sqrt{\frac{(n_1 - 1)s_1^2 + (n_2 - 1)s_2^2}{n_1 + n_2 - 2}\left(\frac{1}{n_1} + \frac{1}{n_2}\right)} \quad \text{with } df = n_1 + n_2 - 2$$

If we can assume that $\sigma_1^2 \ne \sigma_2^2$, then:
the test statistic is   and the maximum error for the confidence interval is

$$t = \frac{\bar{x}_1 - \bar{x}_2}{\sqrt{\dfrac{s_1^2}{n_1} + \dfrac{s_2^2}{n_2}}} \qquad\qquad E = t_{df, \alpha/2} \sqrt{\frac{s_1^2}{n_1} + \frac{s_2^2}{n_2}}$$

with $df =$ the lesser of $n_1 - 1$ and $n_2 - 1$

Hypothesis Test for Two Population Variances (p. 10.9)

The F distribution is used for testing whether or not the variance of one population is equal to that of another. It uses two degrees of freedom, $df_1 = n_1 - 1$ and $df_2 = n_2 - 1$. The test statistic is:

$$F = \frac{s_1^2}{s_2^2} \quad \text{where } s_1^2 \ge s_2^2.$$

Hypothesis Test and Maximum Error for Two Proportions (p. 10.14)

So, the test statistic to be used for comparing two proportions is:

$$z = \frac{(\hat{p}_1 - \hat{p}_2) - (p_1 - p_2)}{\sqrt{\overline{pq}\left(\dfrac{1}{n_1} + \dfrac{1}{n_2}\right)}}$$

!   We state H_0: $p_1 = p_2$. That is, $p_1 - p_2 = 0$ and so can be omitted from the above formula.

The maximum error of the estimate of the population proportion is:

$$E = z_{\alpha/2} \sqrt{\frac{\hat{p}_1 \hat{q}_1}{n_1} + \frac{\hat{p}_2 \hat{q}_2}{n_2}}$$

11. Analysis of Variance (ANOVA)

Use one-factor analysis of variance to test the influence of a single factor on several means instead of just one or two means. Use two-factor analysis of variance to test the influence of two non-interacting factors on the sample data.

One-Factor ANOVA for Equal Sample Sizes (p. 11.1)

For tests involving equal sample sizes, the test statistic uses two estimates of σ in an F ratio:

$$F = \frac{\text{variance between samples}}{\text{variance within samples}} = \frac{n s_{\bar{x}}^2}{(\Sigma s^2)/m}$$

where n = number of values in each sample
   m = number of samples

$s_{\bar{x}}^2$ = variance of the sample means (first estimate of σ^2)

$(\Sigma s^2)/m$ = mean of the sample variances (second estimate of σ^2)

The critical values has degrees of freedom $df_1 = m - 1$ from the numerator of F and degrees of freedom $df_2 = m(n - 1)$ from the denominator of F.

One-Factor ANOVA for Unequal Sample Sizes (p. 11.3)

For tests involving unequal sample sizes, the test statistic uses two estimates of σ in an F ratio:

$$F = \frac{\text{variance between samples}}{\text{variance within samples}} = \frac{\left[\dfrac{\Sigma n_i (\bar{x}_i - \bar{\bar{x}})^2}{m-1} \right]}{\left[\dfrac{\Sigma (n_i - 1)s_i^2}{N - m} \right]}$$

where $\bar{\bar{x}}$ = overall mean (mean of all sample means)
   m = number of samples
   N = total number of data in all samples

The critical value has degrees of freedom $df_1 = m - 1$ from the numerator of F and degrees of freedom $df_2 = N - m$ from the denominator of F.

Two-Factor ANOVA (pp. 11.6–11.8)

Use two factor analysis of variance to test the influence of two non-interacting factors (row factor and column factor) on the sample data. The test statistics use two estimates of σ in an F ratio.

In the following formulas,

$$m = \text{number of rows}$$
$$n = \text{number of columns}$$
$$N = (m-1)(n-1)$$

The first estimate of σ is:

$$\frac{SS_e}{N} = \frac{\sum\limits_{i=1}^{m}\sum\limits_{j=1}^{n}\left(x_{ij} - x_{i\bullet} - x_{\bullet j} + x_{\bullet\bullet}\right)^2}{(n-1)(m-1)}$$

There are two versions of the second estimate of σ. When testing the row factor, use:

$$\frac{SS_r}{(m-1)} = \frac{n\sum\limits_{i=1}^{m}\left(x_{i\bullet} - x_{\bullet\bullet}\right)^2}{(m-1)}$$

When testing the column factor, use:

$$\frac{SS_c}{(n-1)} = \frac{m\sum\limits_{j=1}^{n}\left(x_{\bullet j} - x_{\bullet\bullet}\right)^2}{(n-1)}$$

The test statistics use two estimates of σ in an F ratio. When testing the row factor, use the test statistic:

$$F = \frac{SS_r\big/(m-1)}{SS_e\big/N}$$. The critical value has $df_1 = m-1$ and $df_2 = N$.

When testing the column factor, use the test statistic:

$$F = \frac{SS_c\big/(n-1)}{SS_e\big/N}$$. The critical value has $df_1 = n-1$ and $df_2 = N$.

12. Correlation and Regression

Sample Correlation Coefficient (p. 12.1)

To measure correlation between the two samples, use the sample correlation coefficient (Pearson's Product Moment):

$$r = \frac{n(\Sigma xy) - (\Sigma x)(\Sigma y)}{\sqrt{n(\Sigma x^2) - (\Sigma x)^2}\ \sqrt{n(\Sigma y^2) - (\Sigma y)^2}}$$

where x and y are paired data, and n = number of pairs.

Hypothesis Test for the Strength of r (p. 12.1)

To test the strength of the correlation r between two samples, use a hypothesis test with a t distribution.

Step	Action
1	Formulate the null and alternative hypotheses: H_0: $\rho = 0$ (no significant correlation) and H_1: $\rho \neq 0$ (significant correlation).
2	Determine the critical values of the t distribution using the significance level and $n - 2$ degrees of freedom.
3	Calculate the test statistic $t = r\sqrt{\dfrac{n-2}{1-r^2}}$ and mark it on the sketch of the t distribution.
4	Compare the test statistic to the critical values. With the help of the sketch, draw a conclusion about the null hypothesis.

Regression Line (p. 12.5)

The regression line is one that "best fits" the scatter diagram of the sample paired data. It is used to find y', the predicted value of y. To determine the linear regression equation, follow these steps.

Step	Action
1	Summarize the paired data by finding Σx, Σy, Σx^2, and Σxy. Some of these values might be available from your calculation of r.
2	Calculate the y-intercept $\alpha = \dfrac{(\Sigma y)(\Sigma x^2)-(\Sigma x)(\Sigma xy)}{n(\Sigma x^2)-(\Sigma x)^2}$ and the slope $\beta = \dfrac{n(\Sigma xy)-(\Sigma x)(\Sigma y)}{n(\Sigma x^2)-(\Sigma x)^2}$.
3	The linear regression equation is $y' = \alpha + \beta x$.

Coefficient of Determination (pp. 12.9–12.10)

The coefficient of determination is $r^2 = \dfrac{\Sigma(y'-\bar{y})^2}{\Sigma(y-\bar{y})^2} = 1 - \dfrac{\Sigma(y-y')^2}{\Sigma(y-\bar{y})^2}$.

That is, $r^2 = \dfrac{\text{explained variation}}{\text{total variation}} = 1 - \dfrac{\text{unexplained variation}}{\text{total variation}}$.

Prediction Interval for y (p. 12.11)

To determine the dependability of y', our estimate of y, find the prediction interval for y with these steps.

Step	Action
1	Calculate the standard error of estimate: $s_e = \sqrt{\dfrac{\Sigma y^2 - \alpha \Sigma y - \beta \Sigma xy}{n-2}}$ where α and β are from the linear regression equation.
2	Using s_e, calculate the maximum error: $E = t_{n-2,\alpha/2} \cdot s_e \sqrt{1 + \dfrac{1}{n} + \dfrac{n(x_c - \bar{x})^2}{n(\Sigma x^2)-(\Sigma x)^2}}$ where x_c is a given value of x.
3	Using E, find the prediction interval for y: $y' - E < y < y' + E$.

13. Goodness of Fit Tests and Contingency Tables

Goodness of Fit Tests (pp. 13.1–13.2)

To test whether or not observed data is significantly different from expected data, use a goodness of fit test.

Step	Action
1	Formulate the null and alternative hypotheses: H_0: no significant difference between observed and expected values H_1: significant difference between observed and expected values
2	Determine the critical value of the χ^2 distribution using the significance level and $m - 1$ degrees of freedom.
3	Tabulate the data under these headings. The sum of the last column is the χ^2 test statistic. Class (category) m Observed freq. O Prob. p Expected freq. $E = np$ $O - E$ $(O - E)^2$ $\dfrac{(O - E)^2}{E}$
4	Compare the test statistic to the critical value. Draw a conclusion about the null hypothesis.

Contingency Tables (p. 13.3)

To test whether or not one variable is independent of another, use a contingency table.

Step	Action
1	Formulate the null and alternative hypotheses: H_0: The variables are independent. H_1: The variables are dependent.
2	Determine the critical value of the χ^2 distribution using the significance level and $(r - 1)(c - 1)$ degrees of freedom.
3	Arrange the data in an $r \times c$ contingency table.
4	Calculate the expected values with $E = \dfrac{(\text{row total})(\text{column total})}{\text{grand total}}$.
5	Calculate the test statistic using $\chi^2 = \Sigma \dfrac{(O - E)^2}{E}$.
6	Compare the test statistic to the critical value. Draw a conclusion about the null hypothesis.

Index

Index

A

alternative hypothesis   9.1
analysis of variance
    one-factor   11.1
    two-factor   11.6
approximation rule   6.1

B

bimodal   3.2
binomial
    distribution   5.6, 5.9, 7.5
    experiment   5.6
binomial distribution probabilities table   5.7
binomial probability formula   5.6
boxplot   2.10

C

Central Limit Theorem   7.1
chi-square distribution   8.6
chi-square distribution probabilities table   8.6
class boundary   2.2
class frequency   3.1, 3.5
class interval   2.1, 2.2
class limit
    lower   2.2
    upper   2.2
class mark   2.2, 3.1, 3.5
class midpoint   2.2
class width   2.1, 2.2
coefficient of determination   12.9
column factor   11.6
column sum of squares   11.7
combination   4.8
complement of an event   4.3
conclusion   9.2
conditional probability   4.5
confidence interval   8.1, 8.2, 8.4, 8.6, 8.9, 10.2,
    10.4, 10.7, 10.11, 10.15
continuous random variable   5.1, 7.2
correction for continuity   7.6
correlation   2.13, 12.1, 12.8
    linear   12.1

negative   12.1
no   12.1
positive   12.1
correlation coefficient, critical values table
    for   12.3
critical region   9.1, 9.4
critical value   8.1, 9.1, 9.5, 10.2, 10.9, 10.11,
    10.12, 11.1, 11.3, 11.8, 12.3, 13.1, 13.3
cumulative frequency distribution   2.3, 2.8
cumulative relative frequency distribution   2.4

D

data   1.2
    grouped   2.1, 2.6, 3.5
    ungrouped   2.1, 2.6, 3.4
decile   3.7
decision errors   9.2
deductive reasoning   1.3
degree of confidence   8.1
degrees of freedom   8.3
dependent samples   10.1, 10.2
descriptive statistics   1.2
deviation   3.4, 3.5
    explained   12.9
    total   12.9
    unexplained   12.9
directional test   9.3
discrete random variable   5.1, 7.2
disjoint events   4.4
distribution
    binomial   5.6, 5.9, 7.5
    chi-square   8.6
    cumulative frequency   2.3, 2.8
    cumulative relative frequency   2.4
    frequency   2.3, 2.6, 2.7
    normal   6.1
    probability   5.1
    relative frequency   2.3, 2.6, 2.7, 5.1
    sampling   7.1, 10.14
    standard normal   6.2
    uniform probability   5.2
distributions   2.1
dot notation   11.6

E

English letters, indicating sample statistics   3.6